D1418139

PURCHASING
CONTINUED IMPROVEMENT
THROUGH INTEGRATION

THE BUSINESS ONE IRWIN/APICS LIBRARY OF INTEGRATED MANAGEMENT

Customers and Products

Marketing for the Manufacturer   *J. Paul Peter*

Field Service Management: An Integrated Approach to Increasing Customer Satisfaction   *Arthur V. Hill*

Effective Product Design and Development: How to Cut Lead Time and Increase Customer Satisfaction   *Stephen R. Rosenthal*

Logistics

Integrated Production and Inventory Management: Revitalizing the Manufacturing Enterprise   *Thomas E. Vollmann, William L. Berry, and D. Clay Whybark*

Purchasing: Continued Improvement through Integration   *Joseph Carter*

Integrated Distribution Management: Competing on Customer Service, Time and Cost   *Christopher Gopal and Harold Cypress*

Manufacturing Processes

Integrative Facilities Management   *John M. Burnham*

Integrated Process Design and Development   *Dan L. Shunk*

Integrative Manufacturing: Transforming the Organization through People, Process and Technology   *Scott Flaig*

Support Functions

Managing Information: How Information Systems Impact Organizational Strategy   *Gordon B. Davis and Thomas R. Hoffman*

Managing Human Resources: Integrating People and Business Strategy   *Lloyd Baird*

Managing for Quality: Integrating Quality and Business Strategy   *V. Daniel Hunt*

World-Class Accounting and Finance   *Carol J. McNair*

PURCHASING
CONTINUED IMPROVEMENT
THROUGH INTEGRATION

Joseph R. Carter, D.B.A., C.P.M.
College of Business
Arizona State University

BUSINESS ONE IRWIN
Homewood, Illinois 60430

Sponsoring editor: Jeffrey A. Krames
Project editor: Waivah Clement
Production manager: Ann Cassady
Designer: Larry J. Cope
Art manager: Kim Meriwether
Art Coordinator: Mark Malloy
Typeface: 11/14 Times Roman
Printer: Book Press, Inc.

Library of Congress Cataloging-in-Publication Data

Carter, Joseph R. (Joseph Robert), date.
    Purchasing : continued improvement through integration / Joseph R.
Carter.
        p.   cm.— (The Business One Irwin/APICS library of integrated
resource management)
        ISBN 1-55623-535-6
        1. Industrial procurement—United States—Management.
    2. Materials management—United States.   3. Computer integrated
manufacturing systems—United States.   4. Manufacturing resource
planning—United States.   I. Title.   II. Series
    HD39.C37   1993
    658.7'2—dc20   92-10455

Printed in the United States of America

1 2 3 4 5 6 7 8 9 BP 0 9 8 7 6 5 4 3 2

To my sons,
Roberto and Mario

FOREWORD

Purchasing: Continued Improvement Through Integration is one book in a series that addresses the most critical issue facing manufacturing companies today: integration—the identification and solution of problems that cross organizational company boundaries—and, perhaps more importantly, the continuous search for ways to solve these problems faster and more effectively! The genesis for the series is the commitment to integration made by the American Production and Inventory Control Society (APICS). I attended several brainstorming sessions a few years ago in which the primary topic of discussion was, "What jobs will exist in manufacturing companies in the future—not at the very top of the enterprise and not at the bottom, but in between?" The prognostications included:

- The absolute number of jobs will decrease, as will the layers of management. Manufacturing organizations will adopt flatter organizational forms with less emphasis on hierarchy and less distinction between white collars and blue collars.
- Functional "silos" will become obsolete. The classical functions of marketing, manufacturing, engineering, finance, and personnel will be less important in defining work. More people will take on "project" work focused on continuous improvement of one kind or another.
- Fundamental restructuring, meaning much more than financial restructuring, will become a way of life in manufacturing enterprises. The primary focal points will be a new market-driven emphasis on creating value with customers, as well as greatly increased flexibility, a new business-driven attack on global markets which includes new deployment of information technology, and fundamentally new jobs.
- Work will become more integrated in its orientation. The payoffs will increasingly be made through connections across organizational and company boundaries. Included are customer and vendor partnerships, with an overall focus on improving the value-added chain.

- New measurements that focus on the new strategic directions will be required. Metrics will be developed, similar to the cost of quality metric, that incorporate the most important dimensions of the environment. Similar metrics and semantics will be developed to support the new uses of information technology.
- New "people management" approaches will be developed. Teamwork will be critical to organizational success. Human resource management will become less of a "staff" function and more closely integrated with the basic work.

Many of these prognostications are already a reality. APICS has made the commitment to *leading* the way in all of these change areas. The decision was both courageous and intelligent. There is no future for a professional society not committed to leading-edge education for its members. Based on the Society's past experience with the Certification in Production and Inventory Management (CPIM) program, the natural thrust of APICS was to develop a new certification program focusing on integration. The result, Certification in Integrated Resource Management (CIRM) is a program composed of 13 building block areas which have been combined into four examination modules, as follows:

Customers and products
  Marketing and sales
  Field service
  Product design and development
Manufacturing processes
  Industrial facilities management
  Process design and development
  Manufacturing (production)
Logistics
  Production and inventory control
  Procurement
  Distribution
Support functions
  Total quality management
  Human resources

Finance and accounting
Information systems

As can be seen from this topical list, one objective in the CIRM program is to develop educational breadth. Managers increasingly *must* know the underlying basics in each area of the business: who are the people who work there, what are day-to-day *and* strategic problems. What is state-of-the-art practice, what are the expected improvement areas, and what is happening with technology? This basic breadth of knowledge is an absolute prerequisite to understanding the potential linkages and joint improvements.

But it is the linkages, relationships, and integration that are even more important. Each examination devotes approximately 40 percent of the questions to the connections among the 13 building block areas. In fact, after a candidate has successfully completed the four examination modules, he or she must take a fifth examination (Integrated Enterprise Management), which focuses solely on the interrelationships among all functional areas of an enterprise.

The CIRM program has been the most exciting activity on which I have worked in a professional organization. Increasingly, manufacturing companies face the alternative of either proactive restructuring to deal with today's competitive realities, or just sliding away—giving up market share and industry leadership. Education must play a key role in making the necessary changes. people working in manufacturing companies need to learn many new things and "unlearn" many old ones.

There were very limited educational materials available to support CIRM. There were textbooks in which basic concepts were covered and bits and pieces which dealt with integration, but there simply was no coordinated set of materials available for this program. This has been the job of the CIRM series authors, and it has been my distinct pleasure as series editor to help develop the ideas and facilitate our joint learning. All of us learned a great deal, and I am delighted with every book in the series. But the spirit of continuous improvement is built into the CIRM program and into the book series.

Thomas E. Vollmann
Series Editor

PREFACE

This book focuses on certain aspects of managing the flow of materials through the manufacturing system. The author has selected a limited number of topics which address certain functional area links to purchasing that tend to keep purchasing professionals awake at night. The emphasis is on purchasing and its effect on manufacturing, logistics, and organization. The text was not written primarily for the purchasing professional. As part of the APICS CIRM series of texts, this book is targeted toward non-purchasing managerial personnel interested in ways in which purchasing decisions can impact other functional areas within the firm.

The material presented in this text was obtained from a variety of information sources. These sources included detailed personal and telephone interviews with high-level executives in the areas of purchasing, manufacturing, and logistics as well as case studies of leading-edge manufacturing firms. Information was also excerpted from internal company reports and documents that were made available by many firms over several years. The book is organized into eight chapters dealing with important purchasing links to other organizational functions within the firm. It presents information on various topics such as countertrade, international sourcing, the total quality concept, managing logistics/transportation services, organizational issues, and currency exchange management.

Chapter One stresses the importance of purchasing to the firm by examining the proportion of costs attributable to major types of purchases within all major manufacturing industry categories. The chapter establishes a cost relationship between purchases and labor and demonstrates that purchases are a major component of total cost and profitability, not just for manufacturing, but for the majority of industries.

Chapter Two explores the "purchasing cycle" and its integration within the material planning and control function. The impact upon the purchasing cycle by the new technology called electronic data interchange (EDI) is examined. The chapter describes the cycle, EDI, the potential benefits of EDI, and an implementation process that will increase the likelihood of success for the EDI application.

Chapter Three explores the purchasing and quality linkage through an example of a buying firm and a supplier working together to improve the quality of purchased materials. The chapter does not focus on the mechanics of quality techniques such as statistical process control (spc), but examines the purchasing role in improving quality through better communication between the buying and supplying firms.

Chapter Four examines the purchasing and transportation interface. With the realization of a deregulated transportation environment, the role played by purchasing in selecting transportation services has increased dramatically. Purchasing is demanding both commitment and continuous improvement from transportation suppliers. Transportation relationships are developing into corporate resources which maximize profits, not just cost centers striving to minimize transportation costs.

Chapter Five explores purchasing's role in the receipt of materials and the control of inventory. The primary objective of the chapter is to explain the manner in which the information flow changes as a firm evolves from a manual purchasing system to a computerized purchasing system integrating EDI, bar codes, and production and inventory control.

Chapter Six focuses on international purchasing principles and practices. The emerging importance of international purchasing as a component of corporate strategy is discussed. The details of selecting products/components for offshore buying, source identification and selection, and the mechanics of handling international transactions are presented. The chapter discusses emerging trends in international purchasing.

Chapter Seven explores international countertrade as a component of corporate strategy and an important linkage between marketing and purchasing. Various forms that countertrade can take and their relative advantages and disadvantages are discussed. The chapter provides specific guidelines for effectively using countertrade as a sourcing strategy.

Chapter Eight, the concluding chapter, focuses on the linkage between purchasing and finance. The context for this linkage is the management of foreign exchange rate risk. This chapter explores foreign exchange rate issues in international purchasing and presents a framework of exchange management techniques available to manufacturers purchasing abroad. As firms become more globally oriented in scope and thinking, the topic of exchange rates will increase dramatically in importance within purchasing.

Joseph R. Carter

CONTENTS

CHAPTER 1

PURCHASING AND COMPETITIVE STRATEGY

INTRODUCTION

As U.S. firms strive toward global competitiveness, the effective management of purchases and materials has become a field of great interest and importance. This emerging field will play a critical role in the restructuring and restoration of U.S. industries as they attempt to regain their global prominence. The concept of a domestic market and economy has given way to the more practical realization that all markets are now part of a global economy and American manufacturers must compete as participants in the global marketplace.

It is widely acknowledged by industry leaders that becoming globally competitive requires effective management of the productive and material resources of any organization. Cost, quality, and delivery have become the watchwords in manufacturing operations. In their relentless pursuit to match and exceed foreign competition, U.S. industries are increasingly viewing purchasing not just as a necessary function, but as a weapon in their strategic arsenal.

For U.S. firms, this preoccupation with purchasing is relatively new. The importance of the integrating role that purchasing plays in manufacturing operations deserves greater attention. The principal motivation for this book stemmed from this lack of cohesive information and the realization that the purchasing function will exert a significant influence over manufacturing operations well into the next century.

In many firms, the movement towards an integrated purchasing and materials management strategy has led to pervasive changes in organizational structure, procurement practices, transportation operations, and manufacturing facilities. Clearly, a need exists to bring this important area to the attention of both practitioners and academicians. It is my hope that this book begins to fulfill this need.

This book deals with purchasing and the increasingly important role that it plays in today's manufacturing environment. As U.S. industry struggles to compete effectively in the international market and to regain its global prominence in manufacturing, the restructuring and renewal of manufacturing operations is touted as the principal means to industrial resurgence. This has prompted U.S. industries to examine their organizational structures and those aspects of their manufacturing operations which have the potential to increase competitiveness. Although overused, the term *competitiveness* is the battle cry of many U.S. manufacturing industries.

From studying successful foreign manufacturing firms, U.S. firms have recognized the strategic importance of manufacturing. Many foreign firms have made substantial inroads into traditional American industries such as steel, automobiles, consumer electronics and home appliances. These spectacular gains came at the expense of large U.S. companies. The success of the foreign competition is attributed to their meticulous attention to manufacturing operations, broadly defined to include the procurement of materials and components and the flow of these items into the production facility.

PURCHASING AND MANUFACTURING COMPETITIVENESS

There is no commonly accepted definition of manufacturing competitiveness. It is generally agreed that competitiveness refers to a firm's ability to maintain and enhance its market share. Several generic strategies that a firm can employ to compete successfully in the market are:

- Be a low-cost producer (i.e., concentrate on cost reduction).
- Be a high-quality producer (i.e., concentrate on quality).
- Compete through manufacturing flexibility.
- Or, combine these generic strategies.

The relevance of effectively managing the material resources of an organization to the organization's competitiveness is observed by both practitioners and researchers in operations management. The view of purchasing and materials management taken in this text is somewhat different from that commonly encountered in textbooks on operations management. The conceptual view reflects current industrial practice in stressing **integration across functional responsibilities and the linkages that exist among them** vis-à-vis the materials flow through the manufacturing system. The particular focus of this book is purchasing and the inbound movement of raw materials and components to the production facility.

The three major materials management functions are material purchasing, material transformation, and product distribution. This book focuses on material purchasing and encompasses inbound logistics as well as several of the linkages existing between purchasing and other functional areas within the firm. Purchasing is truly a boundary-spanning function, meaning that purchasing interacts with most other functional areas of the firm on a continual basis. The critical linkage between purchasing and engineering is stressed throughout the book. This linkage is also addressed in the CIRM series text titled *Effective Product Design and Development: How to Cut Lead Time and Increase Customer Satisfaction.*

Purchasing

Purchasing refers to the activities required for obtaining the material resources and services that are needed by the manufacturing system, no matter where in the world those resources exist. Included in this function are identifying sources, selecting sources, developing suppliers, negotiating prices and contracts, developing quality specifications, working with process engineering and product development staffs (simultaneous engineering), coordinating just-in-time purchasing, maintaining the corporate material database, and performing the routine functions of releasing purchase orders, tracking orders, and paying and strategic planning for material resources. Much has been written about these aspects of purchasing, and that information will not be repeated here. The focus of this book is on the strategically important linkages between purchasing and transportation, quality, manufacturing planning, inventory control, finance, marketing, and international sources of supply. The internal workings of the purchasing function, although important, are not within the scope of this text. This text was not written for purchasing professionals, but for others, outside the purchasing area, who may interact with purchasing or are impacted by purchasing decisions on a frequent basis.

Importance of the Purchasing Function

Purchasing is a term used in industry to denote the act of procuring materials, supplies, and services. In a narrow sense, the day-to-day goals of purchasing involve determining the need, selecting the supplier, arriving at an amicable price, issuing the contract, and expediting to ensure proper delivery.

Unfortunately, these goals if taken at face value relegate the purchasing function to a purely clerical status. Nothing could be further from the truth. A broader and more accurate statement of the overall goals of purchasing would include the following seven items.[1]

1. *Provide an uninterrupted flow of materials and services to the operating system.*

It is assumed by many that the basic tenet of purchasing is to obtain the proper materials and services in the right quality, in the right quantity, at the right price, and from the right source. This is not the case. Without an assurance of supply, these other "rights" are meaningless. I once asked a buyer at General Motors what percentage of items that he purchased were custom-designed to meet specific GM needs. His surprising answer was "over 90 percent." If this buyer cannot assure supply, the other factors of quality, price, source, etc., cannot come into play.

2. *Keep inventory investment at a minimum.*

As will be shown later, the purchasing function spends the vast majority of most firms' sales revenues. An easy way to buy is to order in large quantities in order to secure a discounted price and efficiencies of transportation. This creates large inventories of expensive materials.

Such an order policy is unacceptable in the 1990s. The enlightened purchaser continually strives to minimize inventories without loss of price or service quality.

3. *Maximize Quality.*

For years, purchasing has been involved in working with suppliers to improve the quality of incoming materials. But what can purchasing do to maximize a finished product's perceived quality and customer service? Because purchasing is the linkage between the internal and external factory, the function can play a major role in promoting total quality management and assuring customer satisfaction.

4. *Find and develop competent sources of supply.*

The success of any purchasing function can depend on its ability to locate preferred sources wherever in the world they may exist, analyze their capabilities, and select a supplier partner for a long-term relationship. Sometimes a purchaser must look beyond domestic suppliers to find the right choice. The goal of purchasing should be to find sources of supply that will give a competitive advantage to the firm's finished products. More frequently, such sources are discovered internationally.

5. *Standardize, standardize, standardize.*

For years, many firms believed that custom-designed finished goods required custom-designed parts and materials. The Japanese disproved this theory long ago. A major task of purchasing is to gather and disseminate information concerning what standard materials are available that can do the job better and cheaper than parts designed internally. Hopefully, such information is made available during the design stage of product development.

6. *Purchase materials at the lowest total cost of ownership.*

The profit-leverage effect of purchasing throughout the entire materials management cycle can be significant. It has been suggested that price has been replaced by quality as the primary purchase selection characteristic. I hope that this suggestion is untrue. Price, if measured broadly as the total cost of ownership, should always be purchasing's primary material selection criterion.

7. *Foster interfunctional relationships.*

Purchasing buys little for its own use since it is a staff, not a line, function. The staff function exists solely to meet the needs of other functional areas. In that role, purchasing frequently spans the boundaries that separate various functions into departments. This is consistent with the present management philosophy that dictates a cross-functional team approach toward problem solving and process management. Purchasing has proved itself to be a valuable and productive member of these teams.

During the last two decades, purchasing has evolved from being viewed as little more than a clerical function to its present status as an integral part of management. Similarly, purchasing is more than simply buying. Purchasing involves materials and services acquisition, product and process development, capacity planning, and total quality management.

INTERFUNCTIONAL LINKAGES

Purchasing is a boundary-spanning activity. No other functional area develops such close working relationships with other departments within the firm as does purchasing. Advantages in executive management familiarity and overall firm productivity can be gained by encouraging the functional linkages between purchasing and all other aspects of a firm's operations.

The Profit Impact of Purchasing

The procurement function contributes directly to the operating results and profitability of a firm. The fact that purchasing is responsible for spending more than 60 percent of a manufacturing firm's sales dollars highlights the profit-enhancing potential of the purchasing function. Every dollar saved in purchasing is an additional dollar of profit. Figure 1-1 graphically emphasizes this point.[2]

Let's define some key operating performance measures as follows: *profit margin* equals operating revenue divided by sales; *asset turnover rate* equals sales divided by total assets; and return-on-investment (ROI) equals operation income divided by total assets. The performance of top management, and indeed the firm, is often evaluated on the basis of these three financial measures. How can the procurement function impact these performance measures?

Should purchasing be able to reduce the cost of incoming materials by 10 percent, from $500,000 to $450,000, significant changes in these three financial measures would result. Figure 1-2 details these changes graphically. Note that since material purchases decrease by $50,000, average inventories also decrease by half this amount. The profit margin would increase from 5 percent to 10 percent, an increase of 100 percent. The asset turnover ratio would increase from 1.38 to 1.43. ROI would more than double from 6.9 percent to 14.3 percent.

In order to achieve similar results, sales volume would have to increase by a far greater percentage than the 10 percent decrease in material costs. In practice, the cost of materials can easily vary by 10 percent depending upon the skill and expertise of the purchasing team. In addition, such savings in purchasing can be won with little additional investment in the purchasing function.

Purchasing and Marketing

At its basic interface, the purchasing function can assist marketing by buying materials and services at their lowest total cost so that the firm can maintain a competitive position vis-à-vis other firms. But the purchasing function can help the marketing area in other ways. For example, international and domestic countertrade is a neglected area where marketing and purchasing can work together for competitive advantage.

Even though the purchasing function need not be involved early in the countertrade process, some multinational corporations' countertrade arrangements have failed because purchasing was not involved until it was too late. For example, one multinational corporation concluded, without the assistance of purchasing, a $90 million countertrade arrangement with a foreign trading partner. Subsequently, the multinational asked its purchasing function to dispose of the goods taken in this countertrade arrangement. The purchasing function informed management that the goods taken in countertrade were of inferior quality, could not be used internally, and had to be sold at a loss.

This example highlights the strong qualities that purchasing can bring to countertrade arrangements. While the marketing function may agree to almost anything to consummate a sale, purchasing professionals

FIGURE 1–1
Profit Potential of Purchasing

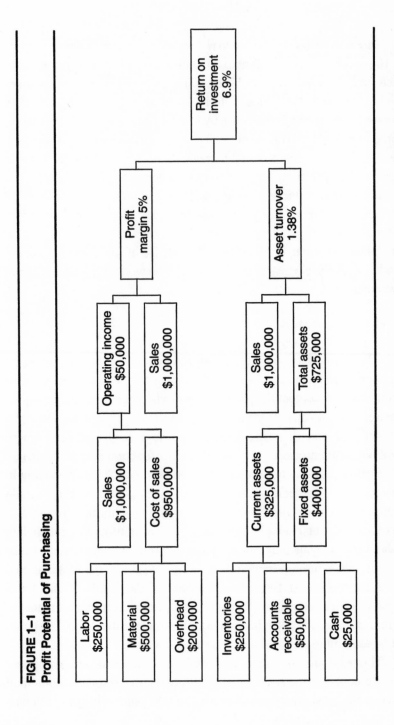

Adapted from Dobler et al., *Purchasing and Materials Management: Text and Cases*, 5th edition (New York: McGraw-Hill, 1990) figure 1–3, p. 13. Reproduced by permission of McGraw-Hill, Inc.

are far more inclined to keep the traditional criteria of price, quality, delivery, and service in mind while evaluating and negotiating counter-trade arrangements. Many insightful companies are adopting a policy that no countertrade arrangements can be made without the involvement and final approval of the purchasing department.

Should purchasing take a more active role in countertrade arrangements, including both developing and negotiating with countertrade partners? The answer is yes for several reasons. First, cost reduction has become a major corporate goal, and countertrade can be a major source of low-cost materials and services. Second, multinationals that find and develop foreign sources of quality materials will have a strategic advantage over their slower competitors who may be stuck with what is left. Third, for the multinational corporation, countertrade is here to stay.

Purchasing and Manufacturing

The procurement function furnishes the materials and services that manufacturing needs to operate. There is clearly a basic common interest between the two functions—so much so, that in many firms the purchasing and manufacturing functions are placed under the same senior managers (e.g., the vice president of Operations, or the vice president of Materials Management if the logistics function is encompassed).

In order to effectively service the needs of manufacturing, purchasing has taken upon itself the goal of assuring an uninterrupted flow of materials from suppliers to manufacturing as its primary goal in most firms. This does not mean that price, quality, service, etc., are unimportant. But if a material or service is unavailable when needed, these other imperatives lose their meaning. It is critical for these two functions to have a complete exchange of information to ensure smooth, efficient control over the materials acquisition process.

Is there a basic difference in the philosophical orientation of the two functions? Manufacturing personnel tend to emphasize protection against stockout conditions (i.e., quantity coordination), while purchasing personnel strive to minimize inventory investments by lowering stock levels or minimizing costs (i.e., price coordination). This basic difference must be continuously controlled in order to reach the proper

FIGURE 1–2
Profit Potential of Purchasing

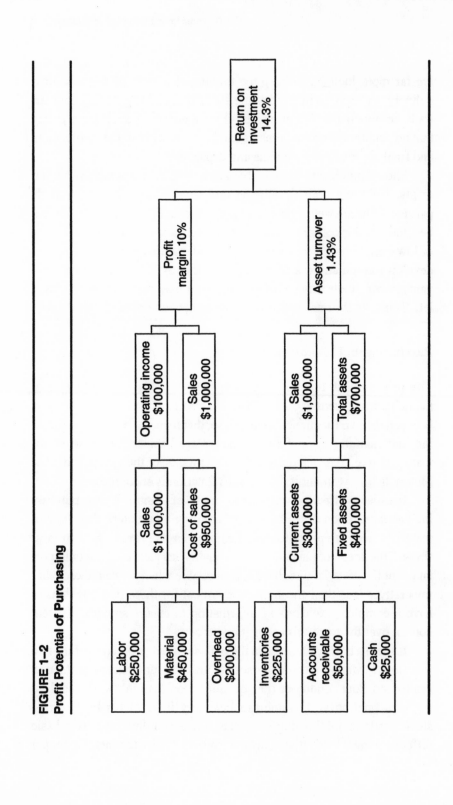

compromise between the two divergent viewpoints. Management should never allow one viewpoint to dominate the other. Both objectives are valid in varying proportions, depending on the strategic value of the resource in question.

If the proper planning and scheduling of supplier capacity is done, lower inventory costs and assured supply can be accomplished jointly. With good timing and a manufacturing resource plan for quantities, efficiencies can be achieved for manufacturing, purchasing and suppliers.

Another issue in price and quantity planning is that of quantity discounts provided by suppliers. When the demand for an item is lumpy, the determination of the quantity to purchase is best decided with discrete lot-sizing methodologies that take into account the supplier's quantity discount price schedule. Quantity discounts should not be a problem for either purchasing or manufacturing, but an opportunity for additional cost efficiencies in operations.

Another area where the two functions can develop friction is that of developing specifications for purchased parts and materials. Manufacturing must understand the capabilities of suppliers before specifying the critical dimensions for materials that may make their cost prohibitive or their availability suspect. Purchasing must constantly search for industry or supplier standard materials that can serve in lieu of custom-specified parts or materials.

Purchasing and Quality Assurance

There has been a major emphasis by the purchasing function in developing quality assurance programs with suppliers. These programs are closely related to the increase in competition from foreign products and the widespread introduction of the just-in-time (JIT) philosophy of manufacture.

Purchasing has discovered that a supplier's quality improvement program can have a significant effect on the price paid for purchased materials. The idea that a firm should pay a premium price for high quality is illogical. The highest-quality supplier invariably has the lowest cost structure and can support minimum prices.

An improvement in the quality of purchased materials can also have a major impact on the internal operations of the firm. As the quality of incoming materials increases, the costs of expediting, production control, inspection, material handling, and indirect labor decrease significantly. The following example of supplier quality improvement at Xerox emphasizes the benefits of such endeavors.[3]

> A quality assurance department was created at Xerox, reporting to the manager of materials management. The department included source surveillance and receiving inspection activities. As a result of the quality improvement program, 50 percent of all incoming material is supplier certified, eliminating the need for receiving inspection at Xerox. Instead, quality assurance personnel visit suppliers on a weekly to monthly basis to discuss a variety of improvement and planning topics.

> The quality assurance personnel work closely with purchasing and buyers located in engineering and design engineers. They join these groups in attending sessions where topics including materials, processes, and tolerances are discussed. In addition, Xerox has instituted a "forward products procurement program" to involve the suppliers in the early stages of the design of new products. Supplier suggestions on the design of new products have resulted in substantial cost savings through improved ease of manufacturing.

Purchasing and Engineering

A major problem in materials management organizations today is the failure to integrate the purchasing function effectively into the engineering system. In many firms, purchasing is not involved early in the requirements development process—for example, during the product development stage. This delay is probably a vestige of the time when purchasing was viewed as a reactive clerical function that issued orders to vendors based on decisions made elsewhere in the organization. If purchasing is to attain its full profit potential within the firm, it must be integrated with engineering early on, during the requirements determination phase of product development.

The responsibility for determining which materials and components to specify for newly developed products is a complex issue because of the conflicting interests, orientations, and biases of the myriad parties that are affected by the intended end item or service. In the traditional (and outdated) model of purchasing and engineering interaction, engineering mostly concerned itself with applying technology to new products and processes, while purchasing concerned itself with attaining the best prices at an acceptable quality for specified items. Not only did these two roles lead to conflict and disrespect between the two functional areas, but optimizing the performance of each function separately frequently minimized the competitiveness of the resulting new product.

In this mode of operation the major supply-oriented tasks of engineering were to provide purchasing with drawings and specifications to be used in supplier development and purchase order generation; to provide engineering change notifications to purchasing and (through purchasing) to suppliers; to provide technical assistance to purchasing during the supplier selection decision; and to analyze materials, parts, and components for purchasing.

In contrast, purchasing's role was viewed as one of a liaison—that is, acting as a go-between for engineering and suppliers; funneling suggested product and process changes from suppliers to engineering; and representing suppliers in their requests for deviations from existing product conformance specifications.

Changes are occurring in this arrangement not because of any in-depth reasoning but because of pure and simple necessity. The old system just didn't work well. Firms are experiencing increasing pressure from both global and domestic competitors. There has been an increased emphasis on total quality and customer satisfaction as a means of competition. Speed to market—that is, reducing the cycle time for product development and manufacture—is becoming the competitive pressure of the 1990s.

There is a new approach toward this purchasing and engineering linkage being pioneered by many astute firms. Under the new model, purchasing and engineering are working cooperatively in cross-functional teams during the product design phase; working together to achieve better make-or-buy decisions; instituting commodity buying teams; and evaluating suppliers in a joint supplier quality improvement effort.

During new product design, purchasing can provide information to engineering concerning the services, components, and materials the firm may decide to buy. Purchasing can assist in establishing price, performance, quality, and reliability targets for product manufacture. It can be a source of information concerning suppliers' abilities to meet the objectives of the new design. Purchasing can also determine the economic and scheduling implications for necessary components, materials, and subassemblies.

With such obvious advantages to linking purchasing and engineering more closely, why hasn't this relationship flourished? There are several reasons why purchasing and engineering haven't been more closely linked, especially during new product design. Many firms have no structured product development process. More specifically, these firms either have not created cross-functional teams or have not empowered those teams effectively. The engineering ethos is to avoid purchasing at all costs and to do everything possible to deal directly with suppliers. Finally, beyond the endemic resistance to change found in any organization, there is frequently a physical separation and resulting lack of communication between purchasing and engineering.

Another way to improve the purchasing and engineering interface is by implementing a commodity team buying approach. Each team would contain representatives from purchasing, material planning, production control, design engineering, and quality. Each team would be responsible for managing the supply base for a specific class of items. Teams would meet regularly to discuss procurement-related issues such as component or supplier selection. In addition, the teams would involve themselves in negotiations, monitor quality performance of suppliers, and provide any requisite supplier reviews or training. This is the approach used by a division of Motorola and explained more fully in the case study at the end of Chapter 3.

What are the benefits of improving the purchasing and engineering interface? At its most fundamental level, such improved communication will foster the transfer of supply market knowledge to engineers while providing needed technical expertise to purchasing. Also, the concept of "design for manufacturability" will be enhanced, manufacturing costs can be reduced, and the time to market for new products should be shortened.

There are several ways to implement this closer relationship. Some companies have used the concept of *co-location*—that is, placing the purchasing staff near or within engineering or vice versa. Other firms have attempted to hire purchasing employees with technical backgrounds, although not necessarily engineers. A very productive approach used by some firms has been to use project teams to develop and introduce new products or at least to formally review all new designs by a cross-functional team that includes purchasing. Finally, some companies have hired "procurement engineers" who work with design engineers on a daily basis to supply information on the commercial implications of different design approaches.

The means used by a firm to foster a closer working relationship between purchasing and engineering is irrelevant as long as it works. The approach used is and should be company-specific. The bottom line is that there are significant gains to be reaped from a closer linking of the two functions.

PROCUREMENT ACTIVITIES

The purchasing cycle includes all of the activities involved in the buying of material and services from the time of recognition of need until the product completes its intended useful life. Types of activities, tasks, and functions on a broad scale include:

- Sourcing: Strategic planning, locating sources of supply, assuring continuity of supply, minimizing risk of supply disruptions, gathering information about supply markets.
- Purchasing: Analyzing decision making: Making make-or-buy decisions, setting standards, certifying suppliers, analyzing values, scheduling, releasing orders, budgeting, planning supplier capacity, and controlling suppliers.
- Contracting: Selecting sources, soliciting bids, doing cost-price analyses, negotiating, establishing relationships with suppliers, evaluating supplier performance.
- Inventory management: Coordinating transportation, receiving, determining lot sizes, controlling purchased inventory, handling materials, disposing of scrap, returning materials.

THE MAGNITUDE OF PURCHASES

Every business and service establishment must purchase. Naturally, there is a wide divergence of the types, sizes and quantities of materials and services purchased by various segments of the economy. Irrespective of the divergence and complexity, competitive demands placed upon business enterprises mandate that the purchasing process be accomplished in a timely and efficient manner.

American industries and governments purchase materials and services costing trillions of dollars annually. The magnitude of these expenditures emphasizes the importance to the economy of the procurement function.

The dollar magnitudes of purchases by type of purchase and by type of industry are presented in Table 1-1. Has there been a failure by American industry to recognize the importance of purchases as a major cost element of business operations? Some would answer in the affirmative.

> Knowledge of cost proportions should direct control efforts. In many companies, although direct labor represents less than 10 percent of the product cost, labor receives a great deal of attention by industrial engineers, factory supervisors, and labor negotiators. Purchased materials may represent 50 percent or more of product costs, but many companies consider purchasing activity routine. . . . [A] shift of attention will benefit the company.[4]

The relationship between business and government purchases and revenues, wages and salaries, and other payments are presented in Table 1-2. A review of nearly all operations-oriented magazines or journals shows that, historically, management has directed an inordinate amount of attention to the reduction of labor costs. In the United States, we have effectively reduced the direct labor force and now desire to do the same to the indirect work force. The data presented in Table 1-2 does not support this disparity of attention between labor and purchase costs. The expenditures for purchases in the 11 industry groups presented in Table 1-2 was, on average, three times greater than the amount spent for wages and salaries.

TABLE 1-1
Business Purchases (millions $)

Industry	Materials and Supplies	Energy	Purchase Category Resale	Buildings and Equipment	Other Purchases	Total
Agriculture	$ 42,958	$ 10,422		$ 27,094	$ 11,185	$ 91,659
Mineral	77,822	7,441	$ 658	49,707	23,974	159,602
Construction	89,891	6,030		13,277	79,473	188,671
Manufacturing	990,084	56,334	52,020	108,383	38,125	1,244,946
Transportation	6,375	16,540		19,329	15,412	57,656
Communication	1,542	1,170		22,525		25,237
Utility	11,074	41,311	89,624	61,873		203,882
Wholesale	5,534	4,063	952,970	24,799	11,615	998,981
Retail	8,585	12,773	720,356	48,950	20,427	811,091
Service	19,521	7,883	7,093	50,447	19,299	104,243
Totals:	$1,300,494	$176,217	$1,870,692	$544,054	$319,582	$4,211,039
Percent	30.9	4.2	44.4	12.9	7.6	100.0

Source: M.E. Heberling, "Purchases by American Businesses and Governments: Types and Dollar Magnitudes," Ph.D. Diss. Michigan State University, (1991)

LESSONS FOR MANUFACTURING

The dollar magnitude of purchases in American industry has significant management implications. Management needs to accord purchasing an even greater level of attention than that which is currently reserved for labor expenditures. Purchases account for the majority of the cost in eight of the eleven industries presented in Tables 1-1 and 1-2. Also, purchases exceed wages and salaries in all industries except services.

Management has developed elaborate systems to control minor elements of total cost while ignoring more important cost elements, such as purchases. Purchases are not just an important cost element in some companies nor is purchasing's importance restricted to manufacturing industries. *Material and services purchases are the dominant cost element in the vast majority of America's industry.* A competitive strategy based on cost reduction is impossible without the involvement of purchasing.

Management should view all purchases collectively and not just from the perspective of the purchasing department. In this light, management should view purchases as major areas of investment. By determining purchase costs in all departments, management can establish goals for cost reductions based on total purchases. By viewing purchase costs in total, they will be in a better position to manage the majority of the company's total costs.

ENDNOTES

1. M. Leenders, H. Fearon, and W. England, *Purchasing and Materials Management*, 9th ed. (Homewood, Ill.: Irwin, 1989), pp. 25-26.
2. Adapted from D. Dobler, D. Burt, and L. Lee, *Purchasing and Materials Management: Text and Cases*, 5th ed. (New York: McGraw-Hill, 1990), p. 13.
3. Example is excerpted from T. Vollmann, W. Berry, and D. Clay Whybark, *Manufacturing Planning and Control Systems*, 2nd ed. (Homewood, Ill.: Irwin, 1989), p. 217.
4. Robert G. Eiler, Walter K. Goletz, and Daniel P. Keegan, "Is Your Cost Accounting Up to Date?" *Harvard Business Review* 60, no. 4 (July-August, 1982), p. 139.

TABLE 1–2
Industry Expenditures

Industry	Expenditures			Cost Ratio
	Purchases	Wages and Salaries	Other Payments	
Agriculture	61.8%	7.3%	30.9%	8.5
Mineral	63.8	14.1	22.1	4.5
Construction	60.8	29.8	9.4	2.0
Manufacturing	63.5	21.6	14.9	2.9
Transportation	51.9	41.9	6.2	1.2
Communication	32.7	32.1	35.1	1.0
Utility	86.0	7.2	6.8	12.0
Wholesale	85.9	7.6	6.4	11.2
Retail	78.1	15.1	6.8	5.2
Service	25.0	39.9	35.2	0.6
Mean	61.4%	20.2%	18.4%	3.0

Cost ratio = Purchases/wages and salaries.

Source: M.E. Heberling, "Purchases by American Businesses and Governments: Types and Dollar Magnitudes," Ph.D. Diss. Michigan State University, (1991)

CHAPTER 2

LINKING BUYER AND SUPPLIER THROUGH ELECTRONIC DATA INTERCHANGE

INTRODUCTION

Electronic data interchange (EDI) is the direct electronic transmission, computer to computer, of standard business forms (e.g., purchase orders, advanced shipping notices, invoices, etc.) between two organizations. In the purchasing environment, the electronic communication takes place between buying and selling firms. Purchasing documents are transmitted "over the wire," eliminating the need to generate hard copies and distribute them manually. The growing popularity of EDI is due largely to the emergence of a broadly accepted standard, (i.e., ANSI X12); the development of relatively inexpensive computer hardware; the growing proliferation of EDI software; and an increasingly competitive purchasing and manufacturing environment.

The importance of EDI is growing dramatically. For example, in the Thursday, July 10, 1986, edition of the *New York Times* under "Technology," an article by Andrew Pollack entitled, "Doing Business by Computer" included the following comments:

> It would be almost unthinkable nowadays for a business not to have a telephone to communicate with customers and suppliers. In the future it may be almost as unthinkable for a business not to have a computer for the same purposes.

"I expect the automotive industry will be a paperless environment in three to five years," said Margaret Goscinski, Associate Director of the Automotive Industry Action Group (AIAG).

"Another factor is that standards are being developed so that all companies and industries can use common electronic order forms and invoices. Before standardization the company might need different computer systems and transmission systems for each customer. Standardizing is the key to making EDI practical" says Edward N. Wong, Production Control Specialist with the Ford Motor Company, which has already eliminated paper purchase orders for auto parts.

Electronic transactions have numerous advantages, proponents say. Today, most of the company's information is stored on computers anyway. When the computer inventory management system indicates its supplies are low, a paper purchase order is made up and mailed. At the other end, the order is received and entered into a computer for billing purposes. It is much simpler and more accurate to have the buyer's computer communicate automatically with the supplier's computer, EDI experts say.

Finally, Input, Inc., in Mountainview, California, a market research firm, estimates that expenditures in the United States in computer service software and transmissions for EDI will grow 100 percent a year from $38 million in 1985 to $1.4 billion in 1990. Still, Input estimates that the number of electronic documents sent will be a small percentage of the tens of millions of business documents sent by mail.

It is imperative that people in organizations who are in a position to influence the design and development of EDI understand what it is and its potential benefits, and follow an implementation process that will increase the likelihood of success for EDI application.

THE PURCHASING CYCLE

A purchasing department buys a variety of materials and services. For the majority of items, the purchasing procedure is a very routine activity. The paper flow in the typical purchasing department is established to support this routine, day-to-day activity, while still providing information to a multitude of different individuals located in several functional

FIGURE 2–1
Manual System

Normal paper flow

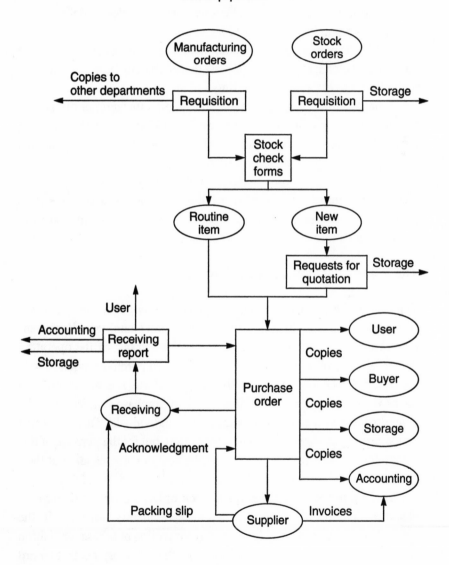

departments. This paper flow has developed over time into a series of procedures that meet three basic requirements. First, the flow of paper permits the efficient use of purchasing resources in conducting the routine activities of the department. Most of the paperwork is done by clerks instead of managers.

Second, the standard flow of documentation is clearly defined. For example, if a purchase order has nine copies, the distribution of each is specified. The reason for this standardization of procedures is so that the multitude of clerks supporting the system can process the documentation with minimum effort and uncertainty.

Finally, the flow of documentation permits managerial discretion. When conditions arise that are not routine or normal, responsible managers are informed about the condition. These managers can take corrective action before any problems arise.

The flow of paper in a manual purchasing system such as the one pictured in Figure 2-1 accomplishes these four objectives. However, this process requires large quantities of paper documents to be moved from and to various locations within and outside the firm.

The Traditional System Paper Flow

The origin of a purchase request frequently occurs either in one of the firm's operating departments or in the production planning and/or inventory control departments. If the need develops in one of the operating departments, such as manufacturing, a standard purchase requisition is generated. This document is strictly for internal use. It is not uncommon for as many as nine copies of this standard purchase requisition to be generated and sent to other departments or individuals. This document contains many pieces of essential information, but for the purpose of this discussion we will deal with only the physical documents and not their content.

If the purchase request is for a stock order, a different document, called a traveling purchase requisition, is frequently utilized. In the manual system pictured in Figure 2-1, the traveling purchase requisition can be a simple indexed card that contains the inventory control record for each item carried in stock. In a manual system, this document is sent to the purchasing department when the stock level for a particular item falls below the item's reorder point.

After these requisitions (nontraveling requisitions) are generated and sent to purchasing, a stock check is usually performed. If the traveling requisition originated in the inventory control department itself, the need for a stock check is eliminated. But for the vast majority of requirements that originate in the operating departments, a stock check is necessary. In some manual purchasing systems, requisitions are automatically routed through the inventory control area before they arrive at purchasing. In firms utilizing computerized production planning and inventory control systems, this entire procedure can be accomplished automatically by computer.

Furthermore, a competitive bidding process is frequently used to select suppliers. When competitive bidding is utilized, the purchasing department sends requests for quotation (RFQ) to a number of potentially capable suppliers of the item in question. It is not uncommon for at least six RFQs to be sent to selected suppliers; therefore, multiple copies of each document are kept on file in purchasing and other internal departments.

Once a particular supplier has been selected, a purchase order is prepared and issued by the purchasing department. At least seven copies of a purchase order are often generated and sent to the various parties impacted by the order. Distribution for a typical purchase order may be as follows:

1. Two copies to the supplier. One copy may be used for purchase order acknowledgment by the supplier.
2. One copy to accounting for comparison with the supplier invoice and receiving report.
3. One copy to receiving for identification and receipt of the incoming shipment.
4. One copy to either manufacturing control or inventory control for planning.
5. One copy retained by purchasing in an open order file, often for expediting purposes.
6. The buyer may retain one copy as a working document.

These seven copies are quite typical and the diagram presented in Figure 2-1 is conservative. In actual practice, changes in company requirements frequently create the need to modify the purchase order. This is accomplished by the use of a change order, which follows many of the same routings as the original purchase order.

There may also be follow-up correspondence between purchaser and supplier after the purchase order is sent and before the order is shipped. The follow-up can be something as simple as a phone call or an official document which is filed. Figure 2-1 also makes no mention of advance shipping notices, which are used by some industrial firms.

When an order reaches the receiving dock, it is accompanied by a packing slip. This packing slip is immediately cross-referenced with the receiving clerk's copy of the purchase order to verify the status of the shipment. For example, the clerk checks to see that quantities and item types are correct. Once shipment has been verified and accepted, the receiving clerk prepares a multiple-copy receiving report. Receiving report copies are usually sent to the following:

1. One copy is retained by receiving for their records.
2. One copy to the using department, in this case either manufacturing control or production planning and/or inventory control, to inform it of material receipt.
3. One copy to accounting for cross-referencing with the supplier's invoice and purchase order.
4. One copy to purchasing to acknowledge completion of the purchase order or to provide back order information.

The normal paper flow shown in Figure 2-1 is quite complex on a manual basis because of the continuous flow and because large quantities of paper are generated and stored. Unless these documents are organized in a systematic manner, much of the information contained on them is useless to purchasing and other departments within the firm.

Table 2-1 summarizes the types and quantities of physical documents that were generated by a surveyed firm's manual system. This function illustrates the impact of paper flow, handling, and storage associated with manual purchasing systems. It should be noted that all of these documents could be eliminated with integrated purchasing—

TABLE 2–1
Manual System Documents

Document	Number generated
1. Internal requisition	9 copies
2. Stock check forms	2 copies
3. Requests for quotation (if new buy)	6 copies
4. Purchase order	7 copies
5. Follow-up correspondence	2 copies
6. Acknowledgement	1 copy
7. Packing slip	1 copy
8. Receiving report	4 copies
9. Invoice	1 copy
Total documents generated	33 copies

manufacturing—supplier systems. Important productivity and competitive gains can be achieved through such systems.

CONTROLLING OVERHEAD COSTS

Much attention has been devoted to reducing direct costs in the manufacture of a product or delivery of a service. Quietly, however, there has been an increasing awareness and anxiety throughout American industry concerning the explosive growth in overhead costs. Recent data suggests that manufacturing overhead averages 35 percent of production costs for American industry.[1] In contrast, the comparable figure for Japanese manufacturers is just 26 percent. For the sake of discussion let's take a simple definition of overhead costs.

Overhead includes all costs related to the manufacture, purchase, and transportation of materials other than direct labor costs, and purchase material costs. Overhead costs include indirect labor, general and administrative expenses, capital equipment costs, engineering costs and logistics costs. Logistics costs can include the movement and coordination of all materials within a firm. Logistics overhead costs include the salaries of all staff personnel that coordinate these movements, as well as the salaries of purchasing, production planning, receiving, transportation, and other staff support personnel. Electronic data interchange can hit hard at these logistics overhead costs.

What are the real causes of logistics overhead costs? The bulk of logistics overhead costs occur because of the sheer volume of transactions that occur in support of the manufacturing system. Such transactions involve exchanges of materials and information that are necessary to support production but that do not directly add value to the finished product. There are transactions that order, execute, and confirm the movement of materials from one location to another (e.g., all order releases, advanced shipping notices, and receipt confirmations transmitted between a buying firm and a supplier). There are transactions that balance the supplies of materials, labor, and capacity (e.g., the conversion of general materials requirements into purchase orders to specific suppliers). There are quality assurance transactions (e.g., the development of specifications and their transmittal to suppliers). There are change transactions which change and update previously transmitted data (e.g., engineering change orders communicated to suppliers). All of the aforementioned transactions cost money in terms of material, time, and personnel resources expended in their support.

Before the advent of EDI, most of the purchasing transactions were handled manually. The cost of processing transactions manually is many times greater than the cost of processing the transaction automatically. EDI can help reduce the cost of processing these transactions in two ways. First, EDI can reduce the cost to process an individual transaction by eliminating clerical input and automating the transmittal process. This should not only reduce the cost of a single transaction, but also increase its effectiveness. The EDI can be so well integrated with the manufacturing planning and control system that data need only be entered once. Second, the movement toward a fully automated EDI system requires an audit of existing transactions and their appropriateness. Such an audit frequently leads to a reduction in the total number of transactions generated. For example, in an EDI environment with advanced shipping notices and real-time receiving reports, are invoices transmitted by suppliers really necessary?

In any case, if a firm cannot greatly decrease the number of transactions, EDI can make each transaction cost less in resource commitment.

EDI: BASIC CONCEPTS AND BENEFITS

In general there are a number of different types of EDI systems. These systems can be classified according to the types and roles of the participants. These include:

- *One-to-many systems*—The focal point is a single company. The other participants are usually the suppliers for the company or their dealers or customers.
- *Value-added network systems*—The center is less defined and the system takes on an electronic warehouse flavor, with many buyers and sellers interacting. It is a logical development of one-to-many systems.
- *Incremental paper trail systems*—Transaction documents are not created at a single point and are routed more or less directly from one company to another. Instead they pass through a chain of intermediaries, each of which adds to the information and documents. It is a logical development from value-added systems.

For example, a single manufacturing firm buying from numerous suppliers and directly connected with those suppliers would be an example of a one-to-many systems. A system with three different manufacturers all buying from the same suppliers through a third-party network would be an example of a value-added network system. Finally, a value-added network system which would link financial institutions and freight forwarders and which would electronically transfer funds and add documentation for export would be an example of an incremental paper trail system.

Benefits of EDI

Numerous important benefits to EDI-using firms have been identified. These benefits are illustrated in Figure 2–2 and include:

FIGURE 2–2
Electronic Data Interchange

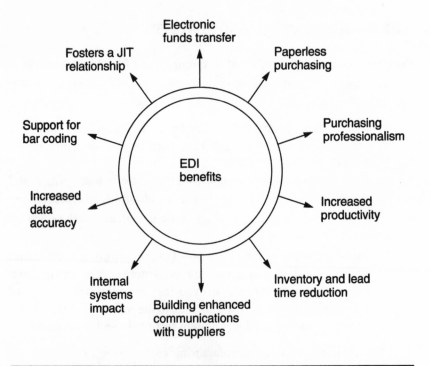

- Increased productivity
- Enhanced purchasing professionalism
- Elimination of paperwork
- Lead time and inventory reduction
- Building enhanced communications and relationships with suppliers
- Support for just-in-time systems
- Support for bar coding
- Increased data accuracy
- Electronic funds transfer
- Establishing the base for further integrated materials and manufacturing systems.

Purchased-item inventory has been reported to have been reduced between 15 and 30 percent. Furthermore, studies have shown that paperwork-related administration in purchasing can take between 20 and 40 percent of the available time. This could be reduced by 50 percent with EDI and other appropriate procurement systems, which could lead to professional contribution improvement.

Also, a number of organizations have reported that staff has been reduced or that additional hiring has been avoided because of EDI. Various organizations have reported a 20 to 60 percent reduction in the cost of transmitting documents electronically versus manually.

Finally, as electronic data interchange applications grow and more experience is established, it becomes appropriate that other types of information beyond the normal transaction documents, such as purchase orders and invoices, may be electronically communicated. This includes quality information, inspection information, statistical process control data, design change data, and so forth. The initial development of EDI for traditional commercial transaction documents could be viewed as just the beginning of improved communications between buyers and suppliers to gain strategic advantages in their competitive processes.

EDI AND MATERIALS MANAGEMENT STRATEGY

Materials management strategy is taking some very dramatic turns so as to better contribute to the overall business objectives of profit, return on investment, market share and market growth. A number of key directions recently being taken include:

- Reducing the number of suppliers and building a longer-term business relationship with a preferred supplier base.
- Closer linking of key suppliers to the design process.
- Implementing just-in-time manufacturing and purchasing systems.
- Linking the buyer/manufacturer with the supplier's manufacturing and delivery process.

- Reducing the number of employees throughout organizations in all functional areas so as to lower indirect costs and overhead and increase organizational flexibility and responsiveness.
- Shortening product development cycle and introduction time of new products.
- Reducing inventory as an asset while still providing required customer service.
- Enhancing supplier quality.
- Improving supplier cost and productivity.
- Worldwide sourcing.

These strategies, directly or indirectly, require improved communication between buyer and seller. This includes both face-to-face and electronic communication between buying and selling organizations. EDI will play an important role in these strategies.

CONTRIBUTION OF EDI TO MATERIALS MANAGEMENT STRATEGY

By utilizing EDI in purchasing, the buyer and supplier are operating practically in a real-time environment which can reduce costly incoming material delays by shortening procurement lead times. Communication becomes more timely, efficient, and accurate. This allows the buyer and supplier to become more market reactive, thereby strengthening their competitive positions.

It has been suggested that the automation of purchasing functions through the use of EDI reduces the total cost of purchases and thereby positively impacts corporate profitability. EDI creates a paperless purchasing environment, reduces administrative and clerical needs, lowers inventory levels, improves data accuracy, and increases purchasing productivity.

All of the previously mentioned benefits are advantages of implementing EDI. What cannot be ignored is the profound impact EDI has on the manufacturing operating system. The EDI technology can be transferred and adopted for use by many functional areas within the organization in addition to purchasing. For example, EDI reduces the lead time (order cycle time) for manufacturing to receive material. It can also provide advance warning to the planning system through the use of electronic order acknowledgements and advanced shipping notices that a vendor is not able to meet its production or delivery schedules. In addition, EDI improves data accuracy because data is entered into the system only once and the system itself can provide many checks and balances. These factors can significantly improve the functioning of the manufacturing planning and control system.

In addition, the EDI logic can support just-in-time manufacturing concepts, automatic identification systems (bar coding) for incoming materials, statistical process control and other efforts designed to more fully integrate the materials management and manufacturing systems. The ripple effect initiated by EDI implementation will be felt throughout the firm for many years.

FUTURE OF EDI

EDI applications have an unbounded future. At the time the telephone was introduced, it seemed like a major technological innovation, which it was. The ways in which people have been able to use the telephone for business purposes since then were unimagined. The same situation holds for EDI.

It is clear that many organizations have already moved in that direction. The transportation industry was a pioneer in developing standards to facilitate transmission of information among railroads and trucking companies. Virtually all waybills used by railroads are electronic. Furthermore, major grocery companies are heavily involved with EDI and use EDI to transmit orders to suppliers.

There are also a number of committees working on standardizing transmission documents. This effort is critically important to the success of EDI. The American National Standards Institute (ANSI) has developed a generic standard for industry, the ANSI X12. Efforts are being made to unite these standards among U.S. and international industries and firms.

Finally, the National Association of Purchasing Management (NAPM), a major trade organization, supports the ANSI X12 standards.

It is clear that organizations of all types should investigate and imple-ment EDI. EDI is the wave of the future and will dramatically and positively change the way in which business is conducted between purchasers and suppliers.

EDI DEVELOPMENT AND IMPLEMENTATION

Implementation of EDI like any planned innovation, is a change process. It will involve different personnel at different organizations. At the smaller firms, top executive and second-level management will probably be involved. At the largest organizations, high-level func-tional managers and possibly third-level executives will directly par-ticipate.

The change process requires that these appropriate individuals un-derstand the current situation, which has been "frozen" for some period of time. They must then decide on the actions that have to be taken to "unfreeze," or change, the current situation. Finally, developing an accepted "refrozen" situation is necessary as people adapt to the new technology, practices, and social relationships. Furthermore, any change process such as the implementation of EDI must be managed. This requires understanding specific actions required by purchasing and materials managers and the change process. Specific major tasks requir-ing management include:

1. Obtaining top management support—the appropriate top management personnel will vary by company as indicated above.
2. Establishing broad-based purchasing and cross-functional sup-port for EDI.
3. Developing a cost/benefit analysis.
4. Auditing all work in every area to see what can be eliminated or supported by EDI.
5. Developing the system.
6. Developing support internally and externally for the ANSI X12 standards.

7. Selecting specific suppliers to participate.
8. Developing necessary legal and auditing actions to support EDI.
9. Establishing internal and external EDI education and training.
10. Deciding on and developing third-party networks.
11. Establishing pilot programs and testing.
12. Monitoring and evaluating the EDI effort.
13. Expanding EDI implementation after review.

CHANGE AND IMPLEMENTATION PROCESS

A planned innovation, such as EDI, requires that attention be given to the management of change in the organization. Personnel with EDI implementation responsibility should be aware of two major considerations which will affect the success of their efforts. First, EDI implementation will affect a number of organization subsystems; secondly, there are established principles that are useful to guide organizational change.

Organizational change is often viewed as the interruption of an existing state (or an unfreezing of a frozen condition). Once a change is made, the impact of the change can be traced to a variety of organizational subsystems. These changes can be seen in (1) the technological subsystem, in which inputs are converted to outputs; (2) the managerial subsystem, in which changes are seen in the way in which managers manage the process and human relationships; (3) the human subsystem, in which human relationships, motivation and expectations are impacted, and (4) the cultural subsystem, in which the organization's views and values may be affected in total.

It is important that purchasing management and other key individuals recognize that EDI implementation will affect people, technology, and structure. Furthermore, systematic management of the effort reduces the likelihood that resistance to change will develop in the purchasing, systems, legal, auditing, finance, and other functional departments, and among suppliers.

There are several principles of organizational change that can be useful to the EDI implementor.

Change is most likely to be accepted and supported if the new values are not too different from prior values—EDI should be presented as a contribution to stated business and functional objectives.

Change efforts are most likely to succeed if they are supported by higher levels and powerful individuals—the support of key leaders should be both informally and formally developed early in the EDI effort.

Those responsible for the change effort should not be easily intimidated by management—individuals charged with EDI implementation should have their own power base, be viewed as high performers within the organization, and be excellent communicators.

Developing voluntary support among those affected by the change will increase the likelihood of success of the EDI effort—significant time and effort should be put forth with both internal personnel and suppliers to educate them about EDI and the benefits of EDI implementation.

The techniques and practices associated with the change must be viewed as acceptable by those people to whom they are applied—EDI should be presented as a continuation of procurement system automation efforts and linked to the need to improve communication with important suppliers.

The organization must be supportive of the type of change being considered—the firm should have a comfort level with using computers to support the purchasing function. Other computer-based applications should already exist.

MEASURING EDI'S IMPACT

What are the tangible benefits of successfully implementing an EDI system? Previous studies have found that the use of EDI is on the rise and that the electronic transmittal of purchase orders is the most common form of EDI communication between buyer and supplier. Despite the limited amount of purchasing information being transmitted electronically, researchers have suggested that substantial benefits have accrued for current EDI users. These individuals state that EDI can significantly impact the cost of doing business with suppliers through a reduction in the costs of purchasing and an increase in purchasing personnel effectiveness. For example, firms that have implemented EDI should be able to reduce clerical costs, clerical errors, and clerical staff because of a reduction in the amount of data that is manually input into the system. In addition, it has been suggested that expediting costs can be reduced and inventories trimmed as ordering lead times are decreased.

From a managerial perspective, several researchers have suggested that when EDI is implemented, the buyer spends less time resolving problems with suppliers because there are fewer errors in the system. This should result in a more effective use of a buyer's time. The buyer should be able to spend more time on productive buying activities (e.g., supplier development, source identification, value analysis, etc.) and less time on purely clerical and expediting activities. The assumption is that the change in the buyer's job content will allow the overall performance of the purchasing department to improve.

None of the EDI studies mentioned above have objectively assessed the expected cost savings resulting from EDI implementation, nor have these studies projected the extent to which the buyer's job will change. The purpose of this last section is twofold: first, to specifically identify the areas within purchasing where costs have been reduced and estimate the magnitude of these reductions resulting from the implementation of EDI: second, to determine the degree of change in time spent by purchasing professionals on several germane buying tasks that resulted from the implementation of EDI.

INVESTIGATION

This section presents the results of a research project that focused on EDI implementation, and involved sending a survey to over 200 purchasing organizations which had previously implemented EDI with a subset of their suppliers. Survey forms were not sent to firms unless a purchasing professional in that firm who was integrally involved in the EDI implementation effort could be identified by name. A significant percentage of the firms were contacted directly by telephone to enhance survey validity. Eighty firms elected to respond to the survey; of these, 54 provided a complete set of responses. These latter respondents were used in the analyses reported in this section.

The degree of EDI implementation was measured on four dimensions: (1) the percent of the firm's supply base involved in an EDI linkage; (2) the percent of the firm's total annual dollar volume of purchases transmitted using the EDI system; (3) the number of different purchasing forms transmitted on EDI; and (4) the length of time since the original EDI implementation. These multiple measurements of implementation were necessary, because a firm might transmit only one type of transaction document (e.g., purchase orders) using EDI to 100 percent of its suppliers and receive few benefits, while another firm might transmit six different purchasing transactions documents using EDI to only 25 percent of its suppliers and receive many benefits.

A survey questionnaire was used to obtain estimates of the extent of actual cost savings and changes in the purchasing professional's work time distribution. This methodology was selected because companies seemingly have difficulty estimating the exact dollar amount of cost savings and because this methodology readily facilitates comparisons between firms of varying sizes, number of suppliers, and dollar volumes of purchases.

Responses

Figure 2-3 graphically presents the results of a survey question asking each respondent to list their particular firm's initial reason for implementing electronic data interchange. The largest percentage of respondents (42 percent) stated that their primary motivation for implementing EDI was to reduce costs. This finding is consistent with many of the previously published articles and books concerning EDI implementation. The more surprising finding was that 21 percent of respondents stated that their firms implemented EDI as a result of customer pressure. Subsequent telephone interviews revealed that these respondents had themselves been previously linked to their customers through EDI and it was a logical step for them to implement EDI with their own suppliers.

FIGURE 2-3
Electronic Data Interchange—Initial Reasons For Using EDI

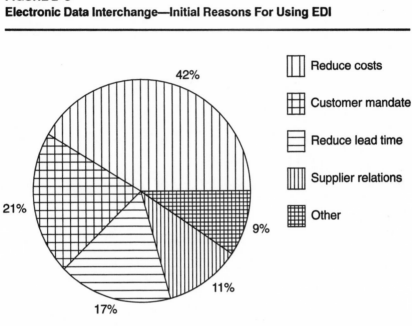

Figure 2-4 graphically presents data describing the EDI system operating experience for the survey respondents at the time each firm answered the questionnaire. This figure shows that 37 percent of the responding firms had implemented EDI less than a year earlier. Another 20 percent had implemented EDI between one and two years previously, while the largest percentage of respondents (43 percent) had at least two years of EDI operating experience. The vast majority of respondents were experienced EDI users with at least one year of system operating experience.

Finally, Figure 2-5 graphically depicts the dispersion of EDI document types that were being transmitted by the respondents to or from their suppliers. By a significant amount, the purchase order was the most frequently transmitted document type, while the request for quote was the least frequent type. Nearly all of the respondents were transmitting more than one type of document. There also was a clear positive correla-

---

FIGURE 2–4
Electronic Data Interchange—System Operating Duration

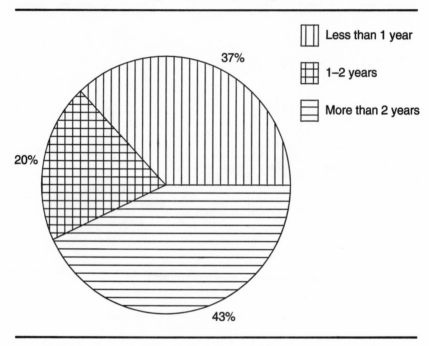

---

tion between length of EDI experience and number of document types transmitted.

Cost Savings and Job Content Changes

One question asked each respondent to indicate the savings that resulted in each of seven areas as a direct result of implementing EDI. The continuum of possible responses for this question ranged from "None" to "Huge." Each respondent could put their individual response anywhere along that continuum. The responses are tabulated in Figure 2-6.

It is apparent from Figure 2-6 that cost savings resulting from the implementation of EDI do not accrue primarily from personnel reductions or new hire avoidance. For personnel reduction, about 90 percent of the firms reported "none" or "little" cost savings and for new hire

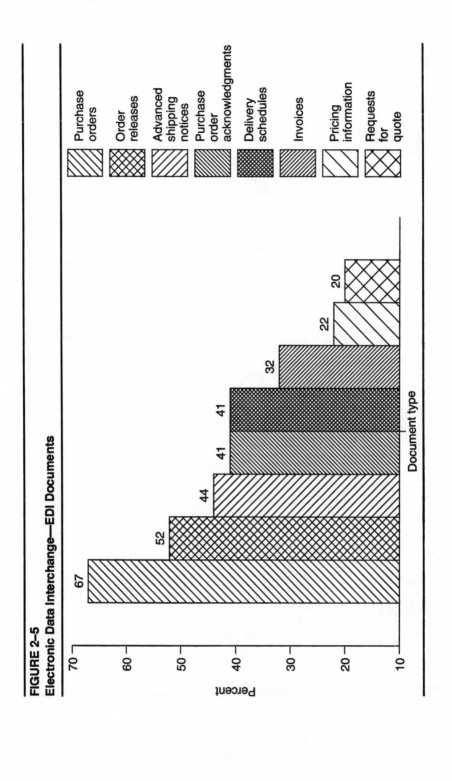

FIGURE 2–5
Electronic Data Interchange—EDI Documents

FIGURE 2–6
Magnitude of Cost Savings

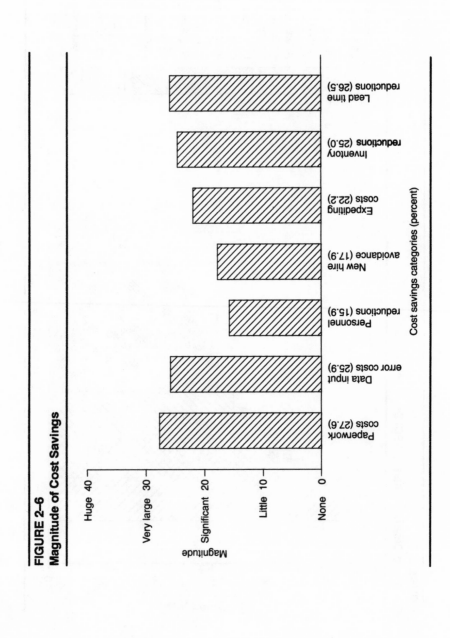

avoidance, about 73 percent of the firms reported "none" or "little" cost savings. This finding is contrary to many previously published articles in trade journals as well as some generally accepted professional opinions. These individuals have consistently touted expected personnel savings as a major reason for implementing EDI. Our survey results paint a different picture.

For our sample of respondents, the greatest cost savings accrued in the areas of paperwork costs, lead time, data input errors, and inventory. As a group, the respondents predominantly reported either significant or very large cost savings in each of these four areas. In addition, the successful functioning of a company's manufacturing planning and control system is integrally impacted by the order cycle time between the firm and its suppliers, the amount of data input errors experienced, and the size and type of inventory.

Figure 2-7 depicts the responses to a question which asked the survey respondents to indicate the degree of change in time spent by purchasing professionals on each of eight buyer-specific tasks. Each respondent was encouraged to place a check on a continuum which ranged from "much less time" through "much more time." The center node on that continuum was designated as "no change." In sum, a check mark to the left of this center point signified that less time was being spent by the purchasing professional on the activity, while a check mark to the right of the center point signified that more time was being spent by the purchasing professional on that activity.

Figure 7 shows that there have been significant changes in the buyers' activities since the EDI system was implemented. For example, it appears that the buyers are spending significantly less time performing routine clerical activities, such as expediting purchase orders, correcting mistakes on orders, tracking purchase orders, and general clerical duties, while spending significantly more time performing professional and managerial activities, such as supplier development and new supplier identification. It is certainly difficult to put a dollar figure on these job content changes, but it does seem logical that the buyers at these firms are becoming more productive. The buyers are spending more time on managerially-oriented activities and less time on clerical duties. The EDI systems seem to be doing what was expected of them in this important area.

FIGURE 2–7
Time Savings

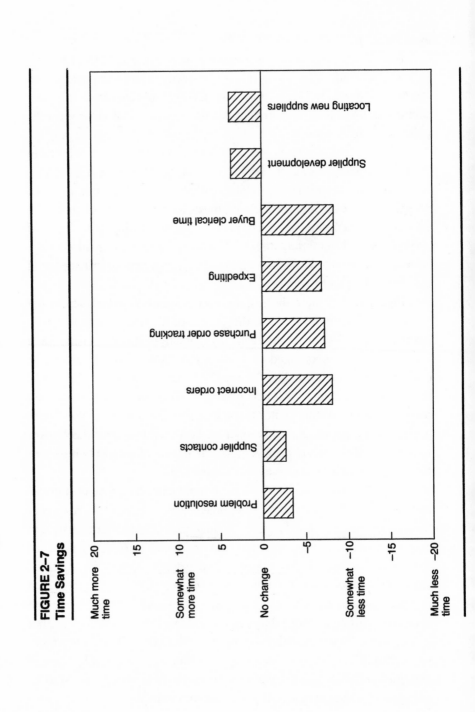

LESSONS FOR MANUFACTURING

The majority of surveyed firms implemented EDI to obtain cost savings. Cost savings did accrue, but such savings did not occur uniformly across all areas. Management should recognize and exploit this variability. It is apparent from Figure 2-3 that the major cost reduction from using EDI does not come from either personnel reductions or new hire avoidance, but reduced paperwork, data input errors, inventories, and lead times. These four areas are critical to the implementation of just-in-time deliveries and a more efficient receiving and material control system.

Once EDI is implemented, buyers spend less time correcting orders, tracking, and expediting purchase orders and more time locating new suppliers and developing existing ones. A valuable and scarce management resource is being conserved. These latter two activities are just the ones a highly skilled purchasing professional should be performing. While routine clerical activities and fire fighting are part of any purchasing job, supplier identification and development have, potentially, the greatest impact on overall firm competitiveness and profitability.

The use of EDI to enhance the communication process strengthens both the buyer and the supplier because their production systems are more synchronized and respond to changes in market demands faster because of the reduced lead times and inventories. Also, a supplier who is communicating with a buyer using EDI can experience similar cost reductions because of the elimination of paperwork and the overhead associated with the paper flow process.

Impact of EDI on Other Functional Areas

EDI implementation programs should be designed to reflect differences between types of organizations, levels of personnel, and functional orientations. In other words, the knowledge and skills necessary to successfully implement and use an EDI system must be identified on an application-by-application basis. For example, the knowledge and skills needed by the auditing function to implement EDI are different than the knowledge and skills needed by the purchasing function. Determining these differential needs is critical to the development of relevant EDI programs.

Table 2-2 provides a knowledge content analysis for various EDI user groups. This figure examines the impact of EDI on various functions within the firm. The left-hand column of Table 2-2 is a list of suggested knowledge requirements which was developed by analyzing training manuals of several firms thought to be leaders in EDI implementation and use. These knowledge requirements, which have been grouped into five categories, are areas of expertise needed to solve the problems and meet the challenges of developing and using an EDI system. The top of Table 2-2 lists affected audiences, divided into three major groups of personnel: suppliers, purchasing and materials management, and interfacing functions. Each of these groups are respectively broken down into a list of specific job classifications. For example, the overall category of suppliers is further classified into key managers, management information systems/users and sales personnel. Table 2-2 cross-references the job categories and the knowledge requirements on the basis of those skills considered to be either "necessary" or "desirable" for effective EDI systems performance.

ENDNOTES

1. Jeffrey G. Miller and Thomas E. Vollmann, "The Hidden Factory," *Harvard Business Review* 65, no. 5 (September–October, 1985), pp. 142–50.

TABLE 2–2
EDI Knowledge Requirements

General Purchasing and Supplier Issues	Supplier			Purchasing and materials management				Interface functions		
	Key Managers	Salespeople	MIS/users	Management	Buyers	Expeditors	Auditing	MIS	Accounts payable	Legal
1. Purchasing in the EDI environment.	X	X		X	X	O	O	X	O	O
2. Organizational changes under EDI.		X		X	X	O	O	O	O	O
3. Task responsibilities with EDI.	O	X	X	X	X	X	X	X	O	O
Electronic Data Interchange										
4. General overview.	X	X	X	X	X	X	X	X	X	X
5. Paperless purchasing.	X	X	O	X	X	O				
6. Computer-to-computer networks.		O	X	X	X	O	X	X	O	
7. Costs and benefits.	X	O	O	X	O		O	O	O	
Transaction Format and Content										
8. Role of standards.	X	X	X	X	X	O	O	X	O	O
9. Standards (ANSI X12).		X		O	O		O	X		
10. Document content.			X	X				X		
11. Data segments and elements.		O	X					X		
Technical Issues										
12. Computer hardware.		O	X					X		
13. Peripheral devices.		X	X	O	X	O	O	X	O	
14. Translation software.			X					X		
15. Application software interfaces.							O	X	O	
16. System security.		X	X	X	X	X	X	X	X	O
17. Communication protocols.		X						X		
Implementation Issues										
18. Implementation tasks.	X	X	X	X	X	O		X		
19. Responsibilities.	X	X	X	X	X	X	X	X	O	O
20. Implementation schedule.	X	X	X	X	O		O	X		
21. Policies, practices, and controls.		X	O	X	X	X	X	X	O	O
22. Troubleshooting.	O	X	X	X	X	O	O	X		

X = Necessary   O = Desirable

CHAPTER 3

LINKING BUYER AND SUPPLIER—INFORMATION CHANNEL DESIGN

INTRODUCTION

A medium-sized manufacturer of specialty steel products changed the wrapping placed around its steel cable during shipping to a particular customer. Unknown to the steel manufacturer, this wrapping, over time, exuded a chemical that caused the exterior of the steel cable to rust, leaving the cable useless for its intended application. It took five weeks before the appropriate individuals in the steel manufacturer's organization learned about the problem and were able to correct it. In the meantime, the customer had production stoppages, large amounts of ruined inventory, and a significant loss of goodwill.

A similar situation occurred when another manufacturer discovered that quality levels on certain purchased components had not improved as expected over time. Upon investigation, the firm discovered that its suppliers, realizing that all material was being inspected carefully by the buyer's quality control department, did very little to improve their processes and quality levels. These quality problems created production shortages, which put pressure on the purchasing department to develop multiple sources and to stress unit cost and variance to standard as the major selection criteria. Uncovering the causes of the quality problems with the suppliers was, unfortunately, a secondary concern.

On the positive side, Ford Motor Company was able to reduce the rejection rate of purchased relays from 40 percent to less than 1 percent within eight months by working closely with some of its suppliers. This raised the net production rate for the supplier from 3,000 relays per line per shift to 6,000. The supplier's manufacturing costs for the relay were also reduced by 20 percent.[1] A similar success story is found in the Xerox Corporation. For five consecutive years, between 1980 and 1985, Xerox was able to reduce its cost of producing copiers by an average of 10 percent per year. Xerox attributes most of these savings to the close working relationships it had with many of its suppliers.

Purchasing people seem to agree these days that better supplier quality management is important competitively. But supplier quality management means more than just reducing defects. Supplier quality impacts each aspect of the buyer/supplier relationship. Major firms like General Electric, General Motors, Honeywell, Xerox, and Ford are introducing a broad variety of programs aimed at working closely with suppliers to improve conformance to specifications, reduce costs, cut inventories, improve designs, reduce order cycle time, and so on. Everybody also seems to agree that the key to achieving these results is better "communication," whatever that may mean, between the buying firm and the supplier.

The purpose of this chapter is to probe beyond the superficial cries for better supplier quality management in order to identify:

- What needs to be communicated to particular suppliers in order to improve supplier quality.
- How to structure information channels with suppliers to support various quality improvement initiatives.
- How to match supplier information needs with channel structure.
- What the implications are for operations managers.

IDENTIFYING INFORMATION NEEDS

Supplier quality management is an important segment of the purchasing function. It is the process by which a buyer works with suppliers over time for mutual gains. These gains can include improved quality, lower costs, and the development and implementation of technical innovations. The actual process of supplier quality management begins after a particular supplier has been selected and purchasing starts releasing orders to that supplier; it ends when the relationship with that supplier is terminated.

Supplier quality management, by its very nature, entails the exchange of large quantities of information between the buyer and a supplier. This information exchange is bidirectional and utilizes some form of communication structure. Table 3–1 provides a partial list of the types of information that might be exchanged between a buyer and a supplier. Clearly, not all suppliers require the same degree of information exchange. For example, when a company buys a generic product or makes a one-time low-value purchase, the information exchanged between the buyer and supplier is limited and the duration of the exchange is short.

When a product is readily available from numerous suppliers or when industry or government standards unambiguously spell out the quality and specifications of the product (e.g., motor oil), a firm can rely on the marketplace to provide much of the needed information and control. Again, the need to exchange information between buyer and supplier is limited.

These two product definitions may apply to a majority of the *items* purchased by a manufacturing firm, but rarely do they apply to a large percentage of the *dollar volume* of purchases. Frequently, the high dollar volume purchases involve those items made to the buyer's own specifications, and they often require special tooling. In this instance, a buyer is purchasing considerably more than just a product from the

---

TABLE 3–1
Types of Information Exchanged

• Order entry	• Manufacturing process
• Price/cost	• Supplier capacity
• Product availability	• Requirement forecast
• Product specifications	• Delivery
• Problems	• Quality assurance
• Financial position	• After-Sales service
• Technical capability	• Product improvement/innovation
• Management capability	

supplier. The buyer is purchasing the supplier's technical expertise and managerial capability. In this case, a premium is placed on the effective and efficient exchange of information.

It seems logical to expect that the best communication structure to use in the exchange of information between a buyer and a supplier may be different for one of the noted scenarios than for the other. Some of the information exchanged for the custom-designed product is routine in nature and similar to the information exchanged between buyer and supplier for a generic product (e.g., order entry, delivery, etc.). But a significant portion of the information that must be exchanged in the case of a custom-designed purchase is unique to that product, that supplier, that application, and that environment. The following section examines the alternative communication structures available to a firm for the exchange of various types of needed information for effective supplier quality management.

ALTERNATIVE CHANNEL STRUCTURES

A systematically designed buyer-supplier communication system is becoming recognized as an essential element of a successful supplier quality management system. The buyer and the supplier have several choices that require careful analysis in each new situation.

Serial Communication System

One common communication system used to provide information and feedback between a buyer and a supplier is shown schematically in Figure 3-1. This communication system provides an indirect flow of information between the using departments in the buying firm and the functional departments in the supplier's organization (with the exception of sales). This model can be called a *serial communication system*.

In the serial system, the purchasing department and the sales department are the focus of all communication between the buying and selling firms. For example, when a problem with purchased materials is detected by a using department in the buyer's firm, it is brought to the

FIGURE 3–1
Serial Communication Structure

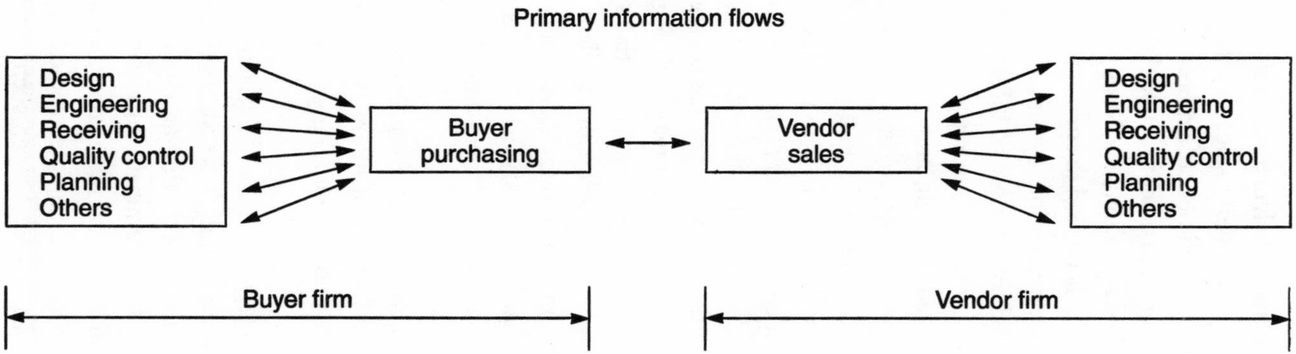

Primary information flows

Design		Design
Engineering		Engineering
Receiving	Vendor	Receiving
Quality control	sales	Quality control
Planning		Planning
Others		Others

Buyer purchasing

Buyer firm

Vendor firm

attention of purchasing, which in turn, notifies the supplier of the quality problem through the supplier's sales department. Unfortunately, such communications are sometimes stifled because interaction between the purchasing department and other functional departments within the firm is not as open and free as it should be. Does purchasing really understand the problem and what is needed to correct it?

The serial communication system provides an opportunity for buyers to control tightly the flow of information to suppliers. It is not uncommon in the United States for company policy to dictate that suppliers be contacted through, and only with the prior approval of, purchasing. This is usually done to ensure that prices and lead times are consistently negotiated and to shield functions, such as engineering and manufacturing, from disruptions. The purchasing and the sales departments are the gatekeepers for their respective firms in the single channel of communication that exists in this system.

The serial communication system has several important advantages. It provides the purchasing department with a great deal of power and leverage in dealing with suppliers. The serial system is used by some firms as a way of playing hardball with suppliers. The theory is that when a buying firm attempts to develop competition among sources, each supplier will work harder to satisfy the customer when no favor, in terms of developing personal acquaintances among using department personnel, is shown to any supplier.

A second advantage is that the serial communication system reduces the potential for confusion that might be created by allowing various functional managers outside purchasing to communicate independently with the suppliers.

Another advantage of the serial system is that it concentrates all communications at one point, thus conserving the buying firm's resources. Most organizations know that an organized chain is a faster and more efficient way to disseminate information than a communication system in which an individual must communicate sequentially with all other interested parties. For example, medical organizations know that it is cost effective to have a gatekeeper, such as a triage nurse, examine a patient's needs before proceeding further with detailed examinations and medical treatment. This cost advantage is maximized when the gatekeeper is knowledgeable enough to communicate effectively, and when there is a premium on one-way—as opposed to two-way—information flows.

Today, when responsiveness and flexibility have become strategic imperatives, the serial communication model has become something of an anachronism. Because in-depth interaction with other functional managers is difficult given the volume of information that purchasing must process, buyers are less likely to become deeply involved in an issue, and much is asked of the purchasing specialist. Taking product design as an example, the purchasing specialist must be knowledgeable in such varied areas as quality control, engineering, and manufacturing. Individuals with these skills are hard to find, train, and keep.

Another disadvantage of the serial communication model stems from the fact that a supplier's sales representatives are unlikely to be the most knowledgeable people in the organization concerning the cause and correction of technical problems that might arise. Other functional personnel may have more knowledge about specific problems of this type. In fact, a supplier's sales department might very well perceive its function as something other than problem correction and thereby give it a lower priority.

Parallel Communication System

Recently, more American firms have begun to move away from the traditional serial pattern of communication between buyer and supplier. Some firms have replaced the serial communication system with one that recognizes the interrelationship between the functional areas in the buyer's organization and their counterparts in the supplier's organization. Such a communication system is characterized in Figure 3–2 and is called a "parallel communication system."

This second system advocates direct communication between functional counterparts in the buying and selling organizations. In the parallel system, buyers and sales representatives play primarily a liaison role. Purchasing and sales are not the sole conduits of information between the two organizations, as they are in the serial system. Rather, these two departments become the coordinators of information transfer.

FIGURE 3–2
Parallel Communication Structure

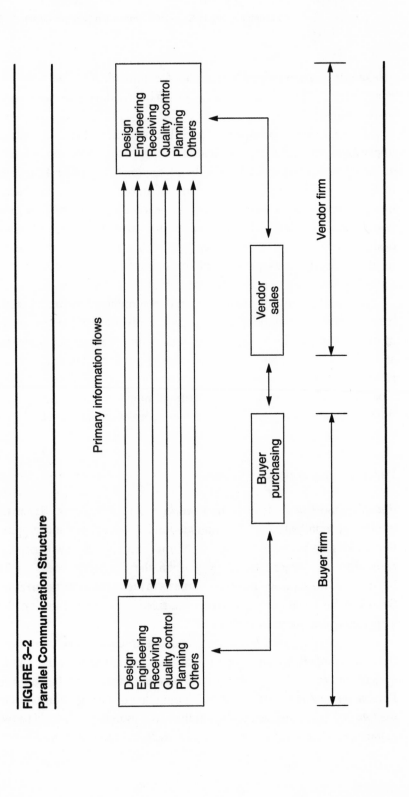

The parallel communication system avoids the basic flaws inherent in the serial system. For example, since information is not transmitted through several organizational layers, the potential for delay and distortion of information is reduced. Appropriate functional personnel from the two organizations that are most knowledgeable about a given issue or set of issues can communicate and work together directly.

The major disadvantage of the parallel communication model is the obvious effort required and difficulty involved in controlling the flow of information between managers from various functional areas in both organizations. This difficulty is exacerbated for a firm that utilizes multiple sources for a single purchased item. In other words, can or should every problem be solved using a functional team approach?

As noted earlier, effective supplier quality management requires a buyer-supplier communication system that provides for the accurate and complete exchange of various types of information. Some of this information is routine in nature, some is not. The nonroutine information rarely is generated or used directly in the buyer's purchasing or supplier's sales departments (e.g., technical design and operating information). In view of this, the parallel communication model may best facilitate a flow of nonroutine information between the relevant functional departments in the two firms. This improved information flow should lead to better performance by both the supplier and the buyer than would be the case with the serial model. Further, it seems logical to expect the utility of the parallel system to increase as the technical difficulty and complexity of issues increase (e.g., the complexity of the supplier's manufacturing task).

CASE STUDY I
IMPROVING SUPPLIER QUALITY AT AMERICAN SEAL

The company that provided the data for this example is a major producer of mechanical seals for a variety of industries. For the purposes of this discussion, the company is called vendor A. Prior to August 1990, vendor A produced its mechanical seals for the aircraft industry in a designated section of the plant, called the aircraft seal (A&S) manufacturing area.

In August 1990, vendor A decided to separate the A&S manufacturing area into two distinct manufacturing areas located in different sections of the plant. One of these areas was dedicated to the manufacture of one particular customer's aircraft seal assemblies. This customer is identified as buyer B. Sales to buyer B composed nearly 30 percent of vendor A's aircraft seal business. Seals for the remainder of the aircraft seal customers were produced in the other manufacturing area. This area continued to be called the AS manufacturing area, even though it was in a different section of the plant. Production of the aircraft seals for buyer B was segregated because buyer B and vendor A decided to work together to implement a statistical process control (SPC) system.

Prior to August 1990, the communication patterns between vendor A and all of its customers for aircraft seals were basically the same. A strict serial communication system was employed. When it became necessary for a customer to communicate with vendor A, the information flow occurred between the buyer's purchasing department and vendor A's sales department. At this time, sales was the sole department that had any form of communication linkage with buyers in customer organizations. Likewise, within the buying organizations, purchasing was the department that controlled the flow of information.

Subsequent to the separation of the buyer B and A&S manufacturing areas in August of 1990, communication between vendor A and all of its aircraft seal customers, except buyer B, remained in the serial mode. However, the communication pattern between vendor A and buyer B changed dramatically. Direct communication between buyer B and vendor A was not limited to the purchasing/sales linkage. Other functional departments, such as manufacturing, quality control, and engineering, were permitted and encouraged to communicate directly with their counterparts in the other firm. Hence, a parallel communication system was developed between buyer B and vendor A.

The parallel patterns of communication that formed between vendor A and buyer B did not develop by accident. They developed as a result of two major programs initiated by the two firms. The program initiated by buyer B was called supplier quality assurance, and was run by buyer B's quality control department. The program was designed to help buyer B's suppliers implement SPC systems. The second major program, initiated by vendor A, involved the transmittal of information concerning designs, configurations, and specifications between vendor A and buyer B. The vehicle utilized for this direct communication was called the design review committee. The design review committee consisted of individuals from manufacturing, engineering, and quality control at vendor A and design engineering, quality control, and purchasing at buyer B. While the primary objective of both programs was to improve the quality of supplier-supplied components, the methods utilized were quite different. The design review committee focused its efforts on defining what quality meant; that is, determining the desired characteristics of the product and developing the specifications for it. In contrast, the SPC effort was aimed at measuring the capability of the processes, ensuring that they could meet specifications, and keeping those processes under statistical control.

Vendor A's separation of the buyer B and A&S manufacturing areas provided an opportunity to analyze, in detail, any differences in the quality performance of the buyer B area (where the parallel communications model was used) and the A&S area (where the serial communication model was used). An initial analysis of the two production areas showed that there was no significant difference between them in terms of type and size of equipment used, or in terms of the experience and skill level of the workers.

The quality assurance data collected in the two manufacturing areas provided an opportunity to analyze the effects of changing the pattern of buyer-supplier communication in both a *pre* versus *post*, and a time series format. The *pre* versus *post* comparison entailed a thorough examination of the average percentage of rejected product per production lot manufactured in each of the two production areas. The *pre* data were taken for a three month period before the parallel communication system was implemented, and the *post* data were taken for a three-month period one year after the parallel communication system was implemented in the buyer B manufacturing area.

Figure 3-3 provides graphic representation of the quality performance analysis. As can be seen, the average percentage of items rejected for the buyer B manufacturing area improved from 30.25 percent in the *pre* period to 15.01 percent in the *post* period. There was no such improvement in the A&S manufacturing area. In fact, the average percentage rejected over the period for the A&S manufacturing area actually increased, although the increase was not significant.

Figure 3-4 shows the ratio of rework hours to direct labor hours over a 14-month period for both the buyer B and A&S manufacturing areas. Parallel communication began for buyer B in August of 1990. This graph indicates that the rework to direct labor hour ratio decreased substantively for the buyer B manufacturing area. This same downward trend is not evident for the A&S manufacturing area.

LESSONS FOR MANUFACTURING

In the anachronistic method of designing products, a product moves from design to engineering to manufacturing engineering to fabrication with little input from outside the firm. The American tradition has been to design first and involve suppliers later. The lack of interaction with suppliers in the design phase can lead to incomplete specifications and controls.

FIGURE 3–3
Vendor A Quality Data

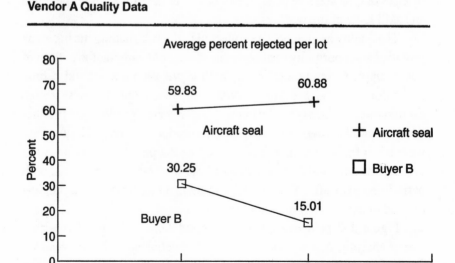

Many firms have seen that this mode of operation has serious drawbacks. A more integrated approach would include design, development, purchasing and key suppliers early in the process. This arrangement allows knowledge gained by purchasing and resident in the supplier base to impact quality up front. Quality can be designed into the product.

If engineering and key suppliers communicate from the beginning of the design stage, costly disputes over critical tolerances can be resolved before manufacture and delivery from the supplier. Once again, design quality in, don't change designs later. Enlightened firms discuss quality and feasibility with their key suppliers at the time of design. These firms analyze the suppliers and processes and work together with the suppliers to achieve their goals for superior product and process quality.

FIGURE 3–4
Vendor A Quality Data

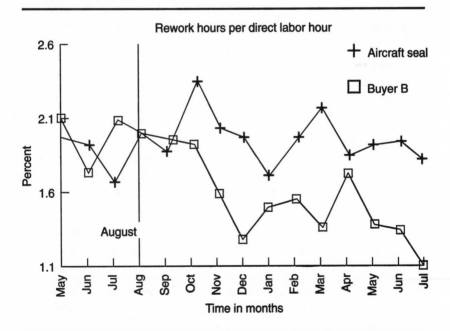

It would not be correct to conclude that a parallel communication system is superior or even more desirable than a serial communication system in *all* circumstances. In this example, however, parallel communication provided a more than adequate means for realizing significant quality improvement. For this reason, and others, firms may implement a parallel communication system to improve supplier performance in a specific area. Nevertheless, it is not practical for most buyers to communicate with every supplier in the parallel mode.

A parallel communication system places a significant drain on the managerial resources of the firm. In today's cost-cutting environment, this could pose problems for many companies. Some firms have used relatively simple approaches to determine when to use each of the two communication systems. One division of a Fortune 500 firm performed an *ABC* analysis on its supplier base, segregating suppliers on the basis of annual dollar purchases. The *A* suppliers were subsequently targeted for parallel communication, while the serial mode was used in most cases with *B* and *C* suppliers. Another firm divided the supplier base by product type. Suppliers producing complex, custom items or those that were deemed to be of strategic importance were considered candidates for parallel communication. Still another firm considered suppliers who supplied components for products that were in the early stage of their life cycle as partners in parallel communication ventures.

The type of communication system utilized impacts the skills and training needed by purchasing professionals. When using the traditional serial system, a buyer clearly needs to be an effective communicator, since the purchasing department is the focal point for communication between buying and selling organizations. Purchasing personnel must be knowledgeable and perceptive, and they must be well-oriented both internally and externally, since they serve as the bridge between their organization and the supplier world. Clearly, they must be reasonably well-grounded with respect to the technical and manufacturing issues associated with their key material responsibilities. When a parallel communication system is used, the technically related requirements for buying personnel typically are somewhat less stringent. Table 3-2 provides a summary of the characteristic differences between serial and parallel communication systems.

TABLE 3–2
Matching Communication Channel Structure with Information Needs

Characteristic	Serial	Parallel
• Product type	Commodity/generic	Customer/strategic
• Stage in life cycle	Maturity	Design/Early introduction
• Information type	Mostly routine	Nonstandard
• Implementation	Easy	Hard
• Vendors participating	Many	Few
• Information Control	Less critical	Critical
• Purchasing personnel		
a. Skills required	Communication	Control
b. Training needed	Broad-based	Narrower

In conclusion, several suggestions are offered for potential users of a parallel communication system. First, some type of control and operating mechanism, such as the design review committee discussed earlier, should be used to minimize communication gaps and to resolve specific problems. Such committees should include functional managers from both the buyer's and the supplier's organizations. The traditional concept of functionally oriented organizations are becoming obsolete. Technology, lead time, and quality requirements mandate a cross-functional team approach to problem solving. Suppliers should be part of these teams. Firms must eliminate the structural barriers that prevent cross-functional and interfirm integration.

Second, a plan should be developed to share with the supplier any dollar savings that are produced by the joint effort. Joint efforts will not occur if all the incentives point the suppliers toward a self-serving position

Third, the users must understand that effective linkages take time to develop because mutual trust must be earned. Firms must explicitly address the internal attitudes that make integration with suppliers difficult. Engineers, purchasing personnel, and supplier personnel must be educated to change their attitude toward working closely together.

Finally, a firm should not wait until a crisis develops. The potential benefits are too great. In times past, slow change was the norm. Today, there is an urgency to creating and sustaining a competitive advantage in manufacturing.

CASE STUDY II
MOTOROLA: GOVERNMENT ELECTRONICS
GROUP (GEG)[3]

In the early 1980s, Motorola, Inc. realized the importance of supplier quality when it saw its market share in the semi-conductor and communications sectors of the electronics industry begin to decline. Japanese competitors were successfully utilizing JIT principles and supplier quality initiatives against U.S. electronics firms. Motorola management studied the problem and familiarized themselves with the Japanese JIT philosophy and Japanese quality and defect level standards. Out of this analysis came Motorola's "six sigma" supplier quality philosophy. Six sigma quality levels equate to .002 parts-per-million (PPM) defective or 99.9999998 percent quality in both part quality and customer service standards.

Motorola GEG's six sigma supplier quality development plan began in 1983 with the announcement of significant reductions of the supplier base. This reduction was the precursor to in-depth investments of time and resources in the remaining suppliers to significantly improve quality. In 1984, GEG centralized the material quality engineering (MQE) function into one division and, at the same time, instituted a formal supplier corrective action system (see Figure 3-5 for a schematic of GEG's quality organization chart). Realizing that a more concerted effort was needed in order to implement such an ambitious program, GEG later decided to manage its supplier base more closely through the formation of a supplier review board (SRB). The SRB was the collection of the MQE, reliability and components engineering, and purchasing functions. The success of the SRB demonstrated the need within GEG for a formal department that specialized in the management of supplier quality. In 1990, GEG combined the MQE, reliability and components engineering, purchasing, incoming inspection, cost analysis, and proposal functions into what was termed the supply management organization (SM) structure (see Figure 6).

At GEG, the manager of the SM organization reports directly to the office of the group general manager (executive vice president). This direct reporting relationship has allowed SM to inject many supplier

FIGURE 3–5
Motorola Inc., Government Electronics Group

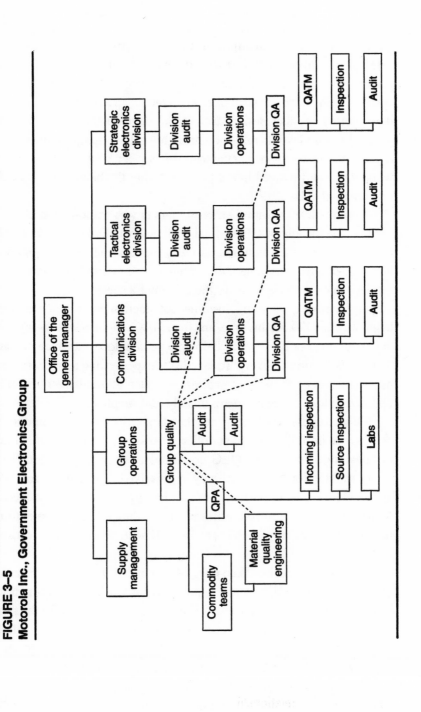

quality initiatives throughout the GEG group. In addition, it demonstrated to each of the group's divisions the emphasis the GEG was placing on supplier quality management. Some of the initiatives pioneered by SM were early supplier involvement, standard parts, and SM input during product design. Before the development of the SM organization, GEG would rely upon its highly skilled design engineers and reliability and component engineers to pioneer new ideas. Suppliers would become involved in the process during the bidding stage of the procurement cycle.

Communication with their customers is very important at GEG. It has become the key of GEG's total customer satisfaction program. The total customer satisfaction program was begun to enhance communication with and responsiveness to customers and to provide products and services that completely satisfy the customer's needs.

Aggressive quality and customer satisfaction goals were set at the beginning of 1990. At this time, and over 3000 suppliers were being used annually, supplier quality was below 90 percent lot acceptance rate and 70 percent on-time delivery. Today, SM is experiencing a lot acceptance rate in excess of 98 percent and 90 percent on-time delivery from an optimized supply base of approximately 800 suppliers.

Other key targets for SM are the standardization of components, continuous improvement of supplier processes, and the establishment of "design-to" suppliers. GEG continues to strive toward total customer satisfaction as evidenced by aggressive key initiatives for 1992 and 1993. These initiatives include having all supplier processes under statistical control, eliminating all inspection of incoming material, and the implementation of EDI with all suppliers, to name just a few.

Incoming inspection is an integral part of SM. Before the formation of SM, an average lot would require over a week to go through the inspection cycle. Today, the average lot take less than a day for the same cycle and the goal is to eliminate inspection entirely. This will be accomplished through certified parts and certified supplier programs— approved manufacturers parts program (AMPP) and distributor inspection certification program (DICP). Each program requires an "approved supplier" status and an excellent quality rating.

FIGURE 3–6
Motorola Electronics Group Supply Management Organization Structure

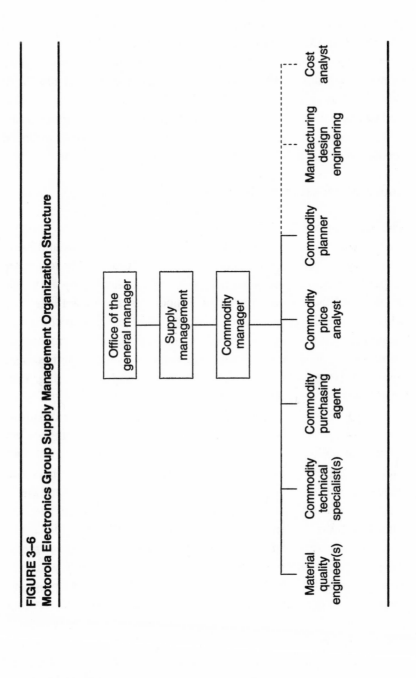

A continuous improvement perspective and philosophy is the key to SM's success. Motorola has embraced the philosophy of continuous improvement throughout the company and is channelling this requirement through the supplier network. The goal is more than just "zero defects." The ultimate goal is to reduce variation in every area. This philosophy has resulted in significantly improved quality, lower costs, reduced waste, shorter development and cycle times, and improved productivity. The inevitable goal is total customer satisfaction.

Supply Management Commodity Teams

The SM organization is comprised of seven commodity teams: passive, connectors, mechanical, microwave, semi-conductor, active, and MRO (maintenance, repair, and operating supplies). Commodity teams are responsible for the daily management of purchasing, quality, and reliability of the various families of components. Each team is responsible for supply base optimization, standard parts, certified parts programs, failure analyses, pricing agreements, supplier corrective actions, and supplier quality surveys. The commodity teams' activities are divided between two area groups, purchasing and engineering. Both groups work toward common goals, but each has specific tasks.

The role of the purchasing group is to administer the buying of components and materials for GEG. Duties include placing orders, monitoring delivery schedules, supporting engineering requests, administering pricing agreements, communicating with suppliers, and generally supporting all program needs. The buyers themselves focus on long-term pricing agreements and supplier interfacing, and are continually working towards reducing the purchasing system cycle time.

Engineering group activities can be separated into two categories: material quality and commodity engineering. A material quality engineer is responsible for supplier quality surveys, certified parts programs, disposition of defective material, and supplier corrective action. A commodity engineer is responsible for part failure analysis, design review assistance, standard parts creation, supporting cost analysis and negotiation support, and supplier problem solving. At GEG, these tasks are combined in several commodity teams and overlap in all of them. The main function of the engineering staff is to provide technical interfacing between the engineering function and suppliers.

Suppliers

Suppliers are evaluated and selected on the basis of technology, quality, financial stability, management attitude, on-time delivery, on-site assessments, responsiveness, and price. No one category is considered significantly more important than the others, but price is considered the first among equals. GEG is a government contractor and government contractors are required to justify their supplier selections on the basis of cost-effectiveness. To date, GEG has not been able to accurately place a figure on total quality and supplier attitude which would satisfy government requirements.

The continuous improvement philosophy is channelled down to suppliers. Each approved supplier must have a continuous improvement plan. Continuous improvement concepts that are expounded to suppliers are total quality management, six sigma capability, process capability, statistical/process control (SPC), implementation methods, individual responsibility, customer/supplier partnerships, and training.

Motorola supports supplier efforts by assisting key suppliers in the implementation of these concepts. Engineers visit suppliers to train supplier personnel and to implement process control evaluations and statistical process control. The commodity teams are responsible for enforcing and instituting continuous improvement plans at suppliers.

Since the formation of the SM organization structure, GEG has developed partnerships with many of its key suppliers. Business forecasts and long-term price agreements assist suppliers in controlling inventory and establishing long-term agreements with *their* suppliers. In addition, suppliers have allowed GEG to examine their processes and provide valuable manufacturing techniques. Improved supplier quality has been the result of these partnerships. When problems do arise, suppliers are willing to work with GEG to implement fast, permanent corrective actions.

Supplier Status

At GEG, suppliers are categorized into three classes—approved, restricted and suspended. *Approved* suppliers are targeted for long-term partnership and recommended for use in new designs. All standard parts must be obtained, whenever possible, from an approved supplier. Approved suppliers participating in the AMPP must be sent a request for quote (RFQ) on all relevant orders. Classification as an approved supplier requires a minimum lot acceptance of 95 percent, as well as a continuous plan for improvement. An acceptable continuous improvement plan includes process control procedures and supporting controls on all critical and major parameters identified in applicable drawings, specifications or purchase orders, a documented training program, management commitment, and a fast response to corrective action requests.

A supplier is classified as *restricted* if the supplier is deemed to have minimal future application within GEG, but is necessary to support present programs. Restricted suppliers require management sign-off prior to the placement of a purchase order. Suppliers may also be placed on restriction for poor quality, nonresponsiveness, and substandard delivery performance. Restricted suppliers are required to have a risk planning assessment visit if orders are in excess of $2,500, lead times are longer than two months, the supplier is on the U.S. mainland, and the quality rating is below 95 percent.

Suppliers are classified as *suspended* when no future business is anticipated. The suspended list prohibits programs from buying materials from these suppliers. Very poor suppliers or suppliers who do not commit to apply for the Malcolm Baldridge Award are usually found on this list.

Certified Supplier Programs

A key aspect of GEG's supplier quality initiative is the implementation of a "dock-to-stock" flow of materials. *Dock-to-stock* means that materials flow directly from a supplier's shipping dock to GEG's production line without incoming inspection or receiving at the company's dock.

To ensure adequate and economical control of material, supplier-responsible defective part information is collected and analyzed to evaluate supplier conformance. GEG has accomplished dock-to-stock flow of material by instituting programs that certify specific parts and part types of selected *approved* suppliers. Entire operations and facilities have been certified.

GEG initially began its certified parts program in 1990 with the creation of the approved manufacturers parts program (AMPP). AMPP was initiated and controlled by the SM function. The SRB reviews and selects candidates for the program. To be considered for AMPP, a supplier must have an approved status, a 100 percent lot quality history on the last 13 lots received for standard parts or 33 lots received for non-standard parts, a 100 percent acceptance rating on the last 500 parts inspected, and statistical process control (SPC) procedures implemented on all critical and major characteristics. SPC data is reviewed quarterly at GEG and meetings are held with the supplier at least once a year. An on-site quality survey is required once every two years to verify the process control program and continuous program plan objectives.

Another dock-to-stock program is the distributor inspection certification program (DICP). As the name implies, the program is targeted towards distributors rather than original manufacturers. The objective of DICP is to eliminate inspection at GEG and perform it at the distributor's facility. DICP candidates must have an approved status, 100 percent acceptance quality rating for the last three months, and the ability to perform inspection at their facility. Other criteria are similar to the AMPP, except that restricted and nonapproved suppliers' parts may be used on DICP.

The benefit for suppliers being on either of these programs is that SM must RFQ them on all orders that relate to their line of business. AMPP suppliers are also designated as "design-to" suppliers and must be considered first for all new design introductions.

CONCLUSION

Supply management (SM) has had a major impact on the way Motorola GEG does business. With the major emphasis leaning towards controlling supplier quality and managing the supply base, supply management will play an even greater role at GEG in the future. The large increases in quality, on-time delivery, and customer satisfaction experienced by GEG have been important for product competitiveness. Improving the last 1 or 2 percent in each category to a truly "zero defects" condition is the challenge that lies ahead for supply management.

ENDNOTES

1. L. Adkins and W. Diller, "Industry's Quiet Revolution," *Dun's Business Month*, June 1983, pp. 72–75.
2. E. Raia, "When Your Supplier Gives You More Than Just an Outside Chance," *Purchasing*, December 1985, p. 45.
3. This case study was adapted from a talk and paper presented by Brett Black and Cindy Fossum at Arizona State University on December 11, 1991.

CHAPTER 4

PURCHASING TRANSPORTATION SERVICES

INTRODUCTION[1]

The role of transportation in the purchasing environment has evolved from being relatively unimportant in purchasing decisions to having a major influence on the contracts signed by today's purchasing agents. Transportation services are developing into corporate resources which generate and maximize profits, instead of corporate costs that need to be minimized.

In the external environment, transportation carriers are becoming logistic service organizations and abandoning the traditional idea of being strictly transporters of products. They are supplementing their traditional transport services with multiple, value-added, tailored services to fit the customer's needs. Several factors have precipitated this change, including industry deregulation, heightened levels of competition, increased demand for flexibility, and enlightened shipper and carrier perspectives of the expanded role of the distribution network.

For a long time, purchasing was playing the game without all of the pieces, which inevitably resulted in something going wrong. Over this period of time, transportation had been purchasing's "missing piece"— the weak link in the logistics chain. Today, all this has changed. Carriers are giving purchasers another weapon for their competitive arsenal.

With sourcing and distribution networks stretching around the globe, effective transportation has become critical. Perceptive purchasing managers are taking control of the manner in which the goods they buy move from source to destination. They are demanding commitment, customer service, and continuous improvement from their transportation service suppliers, and the carriers are responding.

TRANSPORTATION: PURCHASING'S NEW ALLY

With the passage of the Staggers Rail Act and the Motor Carrier Act in 1980, transportation strategy became an important tool for companies to achieve a competitive advantage. Yet many purchasing managers still have not recognized the important changes going on in transportation or understand the significant impact these changes can have on their operations. With so many problems facing companies today, management is asking, "Why should transportation be a high priority on our list of concerns?"

One answer is that transportation is a major cost of doing business. In 1980, U.S. firms spent $204.6 billion, or 7.8 percent of GNP, on transportation. In 1990, U.S. firms spent $331 billion, which was 8.3 percent of GNP. It is estimated that transportation costs will approach 10 percent of GNP by the turn of the century. Physical distribution costs as a percentage of product sales are shown in table 4-1[2].

With distribution costs over 8 percent of sales, it is clear that reducing transportation costs can improve a business' profit performance significantly.

TABLE 4-1
Physical Distribution Costs

Component	Distribution Cost Percent of Sales
Finished goods transportation	3.32%
Warehousing	1.92
Order processing/customer service	0.72
Distribution administration	0.42
Inventory carrying cost at 18 percent	2.18
Other	0.57
Total distribution cost	8.62%

In addition, transportation is important because it directly influences inventory carrying costs, which can be reduced through careful planning. Shipping in smaller quantities can mean savings on inventory; however, the increase in transportation costs may offset this reduction. Premium transportation can cut inventory in transit. The tradeoff between inventory costs and transportation costs must be evaluated. Reliable transportation can help managers reduce safety stocks. Cutting inventory safety stocks, which buffers against uncertainty in demand or in lead times, can reduce costs and help efficiency. Even for a high-value commodity, the extra cost of a fast, reliable carrier may add up to overall savings.

Finally, transportation is central to maintaining a competitive level of service. Time-based competition has arrived. Some customers want a very fast response time, while others prefer scheduled deliveries, even though response time may be longer. Companies that tailor their distribution performance to customers' needs can increase their market share. More specifically, purchasing staffs can use transportation as an offensive weapon when negotiating requirements contracts.

THE ROLE OF TRANSPORTATION INTERMEDIARIES

What is *physical distribution service*? *Physical distribution service* is the interrelated package of activities provided by a carrier or supplier that creates utility of time and place for a buyer or shipper, and insures form utility. From the customer's perspective then, physical distribution service is the mechanism that assures that the goods will be available. Transportation intermediaries are capitalizing on the various utilities of distribution by providing auxiliary services that satisfy the customer.

In the post-deregulation period, transportation intermediaries are becoming known as marketing consultants, transportation organizers, "combiners" of transportation and warehousing, "department stores of transportation services," and middlemen who fit the intermodal pieces together. Intermediaries are emphasizing "customer service" functions rather than simply the traditional transportation efficiency functions as they adapt to the new transportation era.

Transportation intermediaries, or third-party arrangers, are middlemen between shippers and carriers who arrange various transportation and logistical services in a channel of distribution. The economic justification for auxiliary carriers is that they offer shippers lower rates for movement between two locations than would be possible by direct utilization of common carriers. The enabling conditions for auxiliary transport are the existence of minimum charges, surcharges, and less-than-volume rates.

The most common of these middlemen are brokers, shippers' associations, freight forwarders, and shippers' agents. Transportation brokers are defined by the Interstate Commerce Act as persons who hold themselves out to sell, provide, contract, or arrange for transportation services. Typically, the broker will match up a shipper with a certified line-haul carrier.

Freight forwarders are common carriers who assemble and consolidate shipments, provide break-bulk and distribution activities, and utilize another regulated carrier for the line-haul movement. Freight forwarders accept full responsibility for performance on all shipments tendered to them by shippers. They are most common in surface and air cargo. In each case, the freight forwarder consolidates small shipments and then tenders a bulk shipment to a common carrier for transport. At the destination, the freight forwarder splits the bulk shipment into the original shipments. Local delivery may or may not be included in the forwarder's service. The main advantage of the forwarder is a lower rate per hundred-weight and, in most cases, speedier transport of small shipments than would be experienced by direct submission to a common carrier.

Both shippers' associations and shippers' agents perform similar operations to the freight forwarders and brokers except that they are exempt from economic regulation. As the name implies, shippers' associations are voluntary, nonprofit entities with membership centered around a specific industry in which small-shipment purchases are common. Department stores, for example, often participate in shippers' associations, since a large number of different products may be purchased at one location. The office arranges for all members' purchases to be delivered to a local facility. When adequate volume is accumulated, the office arranges for consolidated shipment to the associations' membership base city. Each member is billed on a proportionate basis to traffic moved plus a prorated share of the associations' fixed costs. While a shippers' association is a non-profit organization, a shippers' agent is a for-profit organization which generally concentrates its efforts on consolidation and distribution services for piggyback shipments.

Since deregulation, the number of intermediaries has grown significantly. Prior to deregulation, there were approximately 70 property brokers of regulated commodities in the United States, while after deregulation the number has enlarged to almost 1,000. It is predicted that eventually there will be over 12,000 transportation brokers, giving them control over a major portion of freight movements. It is estimated that brokers control 6 percent of the U.S. domestic general freight movement. Shippers' agents have increased from 200 before deregulation to over 600 after deregulation and shippers associations' numbers have likewise swelled. Each of the customer service functions provided by these intermediaries will be briefly described.

First, the information function focuses on improving the quality of information available to shippers on transportation and logistic price/service options. Second, the matching function is designed to match a shippers' desired level of price/service benefits with logistic capabilities. Third, the selection function focuses on the shipper's difficult decision of selecting a carrier based on customer service requirements. Fourth, expedition and tracing emphasizes routing a shipment in the speediest manner and providing a "trail" to follow the shipment through the system. Fifth, service aggregation creates modal arrangements of service that are consistent with the shipper's service and price preferences. Sixth, the coordination and control function ensures that logistical channel activities are coordinated with transactional channel activities. Seventh, administrative services includes such activities as combining separate bills of lading into one bill, extending credit to shippers or consignees, or simply handling various shipment documentation. Eighth, post-transaction functions include paying freight bills, filing claims with carriers, and resolving disputes between shippers, carriers, and other channel members. All of these services are currently being offered or advertised by intermediaries.

Expanded Role of Carriers

Firms traditionally have viewed transportation carriers as independent entities that provide one service moving products from point A to point B. However, in today's environment, many carriers have become part of their customer's logistics networks. The carriers are able to do this by integrating their strategic and operational expertise with that of their customers and their customers' customers. Actually, the carriers have become logistics service companies capable of performing a variety of distribution functions for their user firms.

Before deregulation, the purchase of transportation was a fairly straightforward corporate decision because of regulated pricing and service controls. However, after deregulation of the major transportation modes, transportation became a commodity to be purchased in much the same way as steel or coal. Shipper firms could select from a variety of carriers, each offering similar service packages, but now at differing price points. Some carriers, not desiring to become part of an undifferentiated market with only price as a competitive variable, have developed strategies to distinguish themselves based on other factors.

For example, Carolina Freight Carriers Corporation provides its customers with a computer network which potentially makes available a variety of services and information capabilities. The network provides price rating, scheduling information, paperless billing through electronic data interchange (EDI), and monthly customer information summary reports of inbound and outbound shipments. In addition, Carolina Freight makes available computer diskettes compatible with IBM PCs which provide rates, bills of lading information and an audit accounts payable system for customers.

As another example, Yellow Freight (a trucking company) allows shippers to access their mainframe computer to retrieve information on shipment status, freight charges, and other timely information. They also bill their customers computer-to-computer.

Truck-leasing companies used to specialize in providing vehicles to private fleets on either a financial or a full-service basis. Since deregulation these companies have aggressively pursued new opportunities and are emerging as some of the largest motor carrier, intermodal, and multimodal companies in transportation. They have expanded their product offerings to include nontransportation services such as warehousing, customer billing, and logistics consulting. Ryder Systems, Inc., for example, one of the nation's largest truck-leasing companies, operates an extensive contract carriage service, a freight management service, and a network of large, irregular-route truckload carriers. Leaseway Transportation Company and its subsidiaries have engaged in vehicle leasing, inventory management, motor contract carriage, a new less than truck load, (LTL) common carriage service, owner-operator truckload transportation, order picking, consolidation activities, warehousing, packaging, and logistics consulting. In contrast to the for-hire motor carriers and railroads, the leasing companies are now positioned to provide either for-hire or single-source leasing services to shippers.

A term often used to describe the provider of fully integrated services is a logistics utility provider. These special service organizations are willing to perform all or any part of logistical requirements as described above. In special situations, they have even agreed to take ownership of selected inventory. Most integrated service firms operate in a consultative manner in that they are problem solvers whose operations are customized to fit their clients' requirements. They have staff and facilities geared exclusively to providing a lowest total cost system.

Shipper and Carrier Benefits

The shippers and carriers have taken advantage of the current situation and are pursuing advantages caused by the new, expanded role of transportation. Overall, shippers have been able to achieve higher levels of productivity. Roadway Express advertises that they can become part of a shipper's JIT plan. The company has an instant shipment tracking system that never loses sight of JIT inventory. Roadway Express developed its complex information system for an automotive assembly plant that was utilizing a JIT production system. Roadway would confirm, via electronic data interchange (EDI), the actual material shipments from suppliers and monitor their progress to meet delivery deadlines.

This system reduced the amount of tracing and expediting of material previously done by the plant, as well as maintained a more accurate control of inbound material flow.

In today's environment, it is essential to remain flexible. The marketplace, systems, and equipment are changing so rapidly that obsolescence has become an important concern. Even if capital is not scarce, companies are realizing that it may be wise to let someone else bear the risk. To illustrate, Leaseway Transportation has offered dedicated, integrated distribution services through its transportation resource management program. In this partnership, Leaseway acquires transportation equipment, facilities, and personnel from the shipper, and returns the distribution service at a higher level of performance. Such outsourcing of distribution operations offers potential for improved cost and service performance, as well as higher-quality labor-management relations skills, and allows the shipper to raise asset performance by repositioning assets to more strategic areas.

Purchasers can also gain greater control over their inbound channel of distribution because of the integration of transportation carriers. This ability to increase control has allowed some purchasers to transfer many logistics functions that they have traditionally performed to others within the logistics network that can do them better.

As an example, Santa Fe Railroad developed a relationship with Ford Motor Company which proved advantageous for both companies. A couple of years ago, Ford invited the railroad to bid on moving auto parts from suppliers on the border with Mexico to an assembly operation in Ontario, Canada. Because the railroad became involved in the logistics planning early, it was able to save Ford considerable expense long before the first shipment was made to Ontario.[3]

Many major manufacturers have developed channel partnerships with logistics service companies. With the volatile merger and acquisition activities of the 1980s, transportation carriers found themselves in an improved market position that enabled them to provide enhanced levels of customer service. Recently, Schneider National and Transco (a Leaseway company) generated substantial revenues from customized services for Mead Paper, 3M, Pillsbury, Dupont, Quaker Oats, and A&P[4].

Bekins High Technology Division is another example of improved, higher levels of customer service. Bekins offers the "Timelock" program in which shippers receive $100 per day, up to the total transportation cost, if shipments are delayed. The company also guarantees pickup and delivery dates from shipments, allows 300-pound minimum weight LTL shipments instead of the industry standard of 500 pounds, and guarantees door-to-door service to all major markets in the United States and Canada for LTL shipments.

The "one-stop shopping" diversification strategy used by carriers also minimizes the carriers' risks associated with carrying a single product and/or service. CSX Corp. provides rail, truck, and barge service, and Consolidated Freightways offers land, air, and sea transportation, import/export assistance, warehousing services, freight forwarding, and logistics consulting services. In today's competitive transportation environment, a company offering a single service is at a great disadvantage. The company with multiple services is much more marketable.

There will always be a need for carriers that choose to compete solely on the basis of the transportation services they provide. The potential trade-offs in this decision are the same as those faced by multi-product firms as they attempt to spread their risks among several products or product lines and create a portfolio. The question of what is best for the overall company must be answered.

A significant mutual benefit that has resulted from the newly diversified environment is the higher levels of coordination between purchasers and carriers. Purchasers have placed a large amount of their traffic out to competitive bid, something unheard-of until passage of the landmark legislation discussed earlier. As a consequence, purchasers are more likely to receive the best service-price package from carriers.

An example of purchaser-carrier cooperation and coordination is the relationship developed between the 3M Company and its carriers. When 3M decided to undertake JIT manufacturing, it developed standards for rating its carriers on their quality performance. First, 3M had to explain what it needed. In June, 1983, it invited executives from the carriers to communicate their goals and upper-management commitments. In the fall of 1983, the company held field meetings during which the carriers, sales representatives, and terminal superintendents received the "nuts and bolts" details of the new strategy. Next, 3M's receiving locations developed local standards such as pickup and delivery times and corporate standards on billing, transit times, and other items. Companywide standards, in place by 1984, were then communicated to the carriers. The coordination was achieved through a give-and-take process involving two-way communication between 3M and its carriers.

The new way of doing business is filled with opportunity for real cost savings and real service improvements. This point was made clear in a recent issue of *Inbound Logistics* magazine.

> Deregulation has enabled the inbound freight buyer to share in the responsibility of developing game plans that reduce costs and improve efficiency. Inbound transportation is not only the least controlled area of transportation, but it is also that one that offers the highest rate of return for the effort invested. . . . Inbound transportation is bargain territory for cost cutting[5].

To the individual firm, the corporate freight bill makes up approximately 10 percent of its expenditures. These are real dollars that com- prise real opportunities for savings. With the rigidity of the rate-making process a thing of the past and a new group of competitive carriers knocking at the door, the potential opportunity is evident.

THE PURCHASING/TRANSPORTATION INTERFACE

To fully exploit the potential benefits, an effective interface or integration of purchasing and transportation must be achieved. Buyers are experiencing expanded roles and responsibilities as they play a greater part in higher-level decision making. Purchasing's effectiveness depends on its ability to adapt to the dynamic conditions of today's market.

A volatile materials management environment requires an ability to plan for and develop improved methods to evaluate supplier organizations and service packages. Knowledge about a more complex set of rules and regulations governing purchases of products as wea. as transportation is essential. How and when to determine what is actually being bought, under what conditions, and at what price are questions that must be addressed. Few absolutes exist. Still, the movement toward centralized materials management is supported by the view that purchasing should not be so much a staff-oriented function as a profit-oriented function. Today, the buyer is becoming more of an overall logistics manager. It seems imperative that purchasing agents understand the expanded purchasing operations.

Given the proper motivation, logistics-oriented suppliers can be of great assistance to purchasing on such matters as supply routes and rates for inbound shipments; routes, rates, and classifications for potential suppliers; rate adjustments; tracing, expediting, and reconsigning shipments; the most economical size of orders from the view of freight charges; pooling of inbound shipments; auditing freight bills; and filing claims for loss and damage. With the increase in international trade, the logistics service suppliers can be of even greater help in regard to shipping and clearance documents, tariff and custom regulations, and various forms of overseas transportation.

An analysis was recently completed on the relative importance of inbound logistics to the purchase decision. The survey data show that on the average the importance buyers placed on logistics services was second only to product quality. It even surpassed price. The results are shown in Table 4-2 below:[6]

These results suggest that inbound logistics service is an important variable in decisions made by today's purchasing managers.

TABLE 4-2
Mean Weighting of Desired Supplier Characteristics

Supplier Characteristics	Mean
Product quality	.176
Logistics service	.171
Price	.161
Supplier management	.152
Distance to supplier	.114
Required order size	.108
Minority/small business	.078
Reciprocity	.046

A New Organization

It is readily recognized that the reporting level within a firm's organization structure can play an important role in the fulfillment of a function's objectives. In the purchasing and transportation functions, the objectives are similar. They focus on cost and the quality of service. With the increased need for productive interaction between these two groups, it is surprising to find that in many organizations they report to different functions at different levels within the organization. With the increased opportunity for economic improvement that can be achieved in the deregulated environment, it seems that the purchasing and transportation functions should operate as closely together and as high in the organizational hierarchy as is feasible.

A good example of this concept is with the Ball Corporation. The corporation is decentralized on the divisional level. In Ball's glass division, which has the heaviest volume of inbound transportation requirements, the purchasing and transportation functions are combined under one vice president of purchasing and transportation. Key advantages of the recent merger were summarized by Ball's manager of corporate purchasing:[7]

1. A reduction in "buck-passing" between some of the subfunctions. People in purchasing are less apt to buy materials where we can't get freight rates to support them.
2. A minimization of conflicting opinions and self-interest.
3. Improvement of vendor relations and reduction of inventory levels.
4. Because of service advantages, the batch materials purchasing manager will not negotiate a new contract without first consulting the transportation manager.
5. The structure encourages more preventative action (proactive). Vendors must consult transportation before charging premium freight costs back to the buyer.

Because of Ball's organizational structure, cross-functionalization is more common and more effective in the current purchasing transportation environment. In many companies, the organization structure is a barrier that impedes the free flow of information, thus limiting communication and coordination.

A Case Illustration

During the past several years, the Ball Corporation has been successful in developing a number of innovative purchasing transportation cost-reduction projects. These projects have produced excellent results that would not have been possible during the previous era of transportation regulation. This case, presented in its entirety by Walters in the Winter, 1988, edition of the *Journal of Purchasing and Materials Management*, will be restated in a condensed version.

Soda ash is a key ingredient in the raw material mix for making glass at one of the Ball manufacturing plants. The primary deposits of this commodity are mined in southwestern Wyoming. Prior to deregulation, the transportation of soda ash from the mine to the plant typically cost several thousand dollars per rail car. Since soda is a dense, bulky commodity, the obvious long-distance transportation advantage lies with the railroad. As competition in the glass-making business became more intense, the cost of transporting soda ash became a rather interesting issue. There was only one rail carrier geographically positioned to service soda ash purchasers, and that carrier refused to negotiate or cooperate to any extent with respect to service or rate modifications. The competition, restrained by regulation, was no threat to the rail carrier's soda ash business.

Following deregulation, however, the potential competitive situation changed drastically. Ball's purchasing and transportation people worked together to explore various alternative solutions. Eventually, they were able to create a unique arrangement with the soda ash supplier and a competing railroad located 200 hundred miles away. The result was true competition and low costs.The solution involved an intermodal transportation arrangement. A contractual arrangement was negotiated with a motor carrier to haul the soda ash from the mine to a newly renovated transfer station approximately 200 miles from the second rail line. From the transfer site, the new rail carrier transports the soda ash to Ball's plant in the East. In the final analysis, the cost savings produced by this new competitive arrangement netted out at approximately $7 per ton of soda ash. The annual volume purchased by Ball generally runs around 18,000 tons, producing savings of over $126,000 per year at this one plant location.

This case illustrates what can be done when the integral components of a process—in this case purchasing and traffic—interact cooperatively and creatively to develop an approach for solving a routine problem.

Ball Corporation attributes much of its success in the past to an intracorporate strategic-planning process. Each staff unit is required to present to management a five-year detailed operating plan. All costs, both current and anticipated, are analyzed. Purchasing and transportation work closely together in the forecasting process and eventually in presenting the data. It becomes a necessity to integrate these two functions in order to produce accurate strategic plans and forecasts.

One recent survey reinforced the impact of corporate strategy on transportation and logistics activities. The survey found that "changes in corporate strategy" and "changes in company size, complexity, or share of market" are the factors most likely to cause significant changes in the organizational structure of logistics and transportation activities[8]. Again, the issue is that functions may be willing to communicate with one another, but the organizational structure may not be conducive to such activity.

Key Elements

Communication, interaction, and perhaps integration seem to be the key elements in the development of an effective interface between the purchasing and transportation functions. It is obvious that a formal, as well as an informal, set of communication linkages should be established within the organization to ensure the necessary interaction that will likely optimize the firm's operating position. If there are few lines of communication and interaction between the various functions, significant inefficiencies are likely to arise.

Today, the interdependency of functions, specifically purchasing and transportation, is stronger than ever. Recognition of this interdependence is critical for mutual advantage to result. Cross-functional, integrated organizations with formal lines of communication are in the forefront of efficient operations. Therefore, commitment to better and more sensitive communication is essential.

The use of internal, computer-based communications is becoming commonplace. Many firms have purchasing and transportation linked directly together with a simple computer network. In some organizations, it is possible for a buyer to obtain optimal shipment routings simply by entering the description, destination, origin, and weight into the system. The transportation department keeps it updated as required by changing conditions in the transportation market.

Purchasing can be helpful to their transportation counterparts in negotiation and pricing. For purchasing people, negotiation and pricing have been a fact of life. Today, cost determination is often attached to negotiation, and the resultant pricing levels are a reflection of this activity. In many cases, both the buyer and the transportation specialist have contributions to make to the negotiation process. In general, a team approach should be used in negotiation. The team should usually accomplish more together than is possible individually.

A final point is that the communication linkages should not stop at the internal level. The same commitment to timely, accurate information to and from suppliers is just as important. Sellers are penalized when buyers do not keep them informed about their current and proposed production and/or supply requirements. Buyers, on the other hand, lose time and money when suppliers do not deliver.

LESSONS FOR MANUFACTURING

It is becoming increasingly important for companies to develop partnerships or strategic alliances within the logistics network. Strategic alliances are business relationships in which two or more independent organizations decide to work closely together to achieve a set of specific objectives. Although some alliances may not be beneficial in the short run, it is believed that in the long run both parties will profit.

For purchasing, it is critical to control transportation costs. Many purchasers buy their goods in a package deal that includes transportation cost. Frequently, the cost of transportation gets buried in the price. There is great opportunity for the generation of cost savings and quality of service improvements if the purchasing and transportation functions can interface effectively within the buying organization. Deregulation has provided the space for this activity.

One of the less observed results of transportation deregulation has been the explosive growth of transportation intermediaries or third-party specialists such as brokers, shippers' agents, and integrated leasing companies. The expanded availability of efficiency and service quality of transportation intermediaries after deregulation means that opportunities now exist for a closer correspondence between traditional purchasing and logistics concerns. The use of these intermediaries by industrial purchasers as an outsourcing strategy has the potential to both reduce delivered product costs and enhance service quality for a company's competitive advantage. In a broader sense, intermediaries may be ideally positioned to assist in coordinating and processing information for the entire value-added chain. For the current general business environment of increased time-based competition, intermediary functions may be particularly well suited for facilitating both purchasing efficiency and effectiveness.The emphasis of this chapter has been on the rapidly changing role of purchasing in terms of responsibilities, scope, and importance and the concept of a purchasing/transportation interface. Fortunately, a healthy purchasing department brings substantial benefits to those individuals and organizations that are willing and able to coordinate other departments' efforts with it. For the logistics manager, it provides a unique opportunity to reduce costs while doing a better job of controlling inventory, scheduling inbound transportation, and handling related information flows more efficiently.

Purchasers are changing their philosophy to reflect current conditions and provide direction performing their purchasing functions and in relating to logistics managers. They are focusing on suppliers rather than supplies. Ironically, this has led to improved product quality and vendor performance.

However, the firm's main goal is to provide customer satisfaction. Firms must begin to realize that satisfying the customer means the creation of quality through the addition of form, possession, time, place, and quantity utilities to their products. Form utility is created by engineering and manufacturing; possession utility is added by marketing; time, place, and quantity utilities are created by the logistics activities.

With the addition of quality in time, place, and quantity utilities to the firm's output, firms will strengthen their competitive position in today's global markets. Accordingly, both purchasing and logistics must perform their respective functions in order to solve their part of the customer satisfaction equation.

ENDNOTES

1. The author gratefully acknowledges the assistance of Scott Nemeth, my graduate assistant, in the construction of this chapter.

2. Lewis M. Schneider, "New Era in Transportation Strategy,"*Harvard Business Review,* March–April 1985, pp. 118–26.

3. Ira Rosenfeld, "Despite Growth, Contract Logistics Still the Exception and Not the Rule," *Traffic World,* October 7, 1991, p. 11.

4. James Stock, "The Maturing of Transportation: An Expanded Role for Freight Carriers," *Journal of Business Logistics* 9, pp. 15–31.

5. James Curley, "Consignees Call the Signals for Transportation Savings," *Thomas Register's Inbound Logistics,* October 1985, pp. 13–15.

6. William Perreault, "Physical Distribution Service in Industrial Purchase Decisions," *Journal of Marketing* 40, pp. 3–10.

7. Peter Walters, "The Purchasing Interface with Transportation," *Journal of Purchasing and Materials Management,* Winter 1988, pp. 21–25.

8. Jack Farrell, "Department Organization—Forces for Change," *Traffic Management,* March 1989, pp. 67–77.

CHAPTER 5

SUPPLIER BAR CODES: CLOSING THE EDI LOOP

INTRODUCTION

The use of computers by purchasing personnel has grown rapidly over the last decade. Surveys of computer use by purchasing managers show that much of their use is for routine purposes, such as inventory control and purchase order generation. Two critical issues in the use of the computer for these types of applications are (1) the timeliness of the information, and (2) the accuracy of the information. If the purchasing department is to function effectively, it simply must have accurate and timely information.

The speed and accuracy of data collection and dissemination can be improved by reducing the human element in the process. One set of tools for reducing the human element in this process is the electronic transmission of data between a buyer and seller, or electronic data interchange (EDI).

The introduction of EDI can significantly impact the flow of information through the purchasing cycle. This changed information flow will affect the paper documents traditionally used to transmit necessary purchasing information between a buying firm and its suppliers. A fully functioning EDI system can eliminate nearly all external paper. A popular phrase coined to describe that elimination of paper is "paperless purchasing." This phrase does not do justice to the work flow changes that occur as EDI is placed in operation. Paper does not just disappear in an EDI environment. As more and more transactions are processed electronically, more and more of the paper flowing between a supplier and buyer is eliminated.

Bar coding is a natural complement for electronic data interchange systems. Bar coding closes the electronic loop between the buying and supplying firms. Where EDI can eliminate most external purchasing documents, bar coding can be used to integrate the receiving function electronically with the computerized purchasing, materials management, and accounts payable systems.

The primary objective of this chapter is to explain to the reader the way in which the paper flow changes as a firm evolves from a manual system to a computerized purchasing system to a computerized purchasing system integrating EDI and supplier bar codes. Just as a firm must evolve from a manual to an integrated computerized purchasing system, the physical transaction documents must disappear in an evolutionary manner over an extended period of time.

The first section of the chapter details the receiving function and an EDI system linkage between a buyer and supplier. This section describes the evolutionary elimination of paper documents as a purchasing system evolves from being manual to computerized. The second section examines the bar coding of incoming materials in an EDI environment and provides two case examples of a bar code and EDI interface. It is the opinion of the author that the inherent potential of bar coding can be realized more fully in an EDI environment. The final section of the chapter presents a summary of the discussion and implications for managers in the process of integrating a supplier bar code and EDI system.

THE RECEIVING FUNCTION

When an order reaches the receiving dock, it is accompanied by a packing slip. This packing slip is cross-referenced with the receiving clerk's copy of the purchase order to verify the status of the shipment. For example, the clerk checks to see that quantities and item types are correct. Once shipment has been verified and accepted, the receiving clerk prepares a multiple-copy receiving report. Copies of the receiving report are usually sent to the following areas:

1. One copy is retained by receiving for its records.

2. One copy is sent to the user department to inform it of receipt.
3. One copy is distributed to accounting for cross-referencing with the supplier's invoice and the purchase order.
4. One copy is delivered to purchasing to acknowledge completion of the purchase order or to provide back-order information, if necessary.

The normal paper flow of the receiving activity can be quite complex because of the continuous document flow and because of the large quantities of paper generated and stored. Unless these documents are organized in a systematic manner, much of the information contained is useless to purchasing and other departments within the firm.[1]

Computerized Purchasing Systems and the Evolution to EDI

Electronic data interchange does not eliminate internal documentation. EDI has been defined as the exchange of business information between firms in machine-readable form. EDI was developed to eliminate the *external* documentation between a supplier and a buyer, not the *internal* paper. A computerized purchasing information system, with or without EDI is needed to eliminate internal documentation. Many firms have computerized purchasing information systems without EDI that are quite sophisticated. By and large, these systems have eliminated much of the internal paper documentation through intelligent design, without the use of EDI.

EDI is only part of an integrated purchasing management information system. Since EDI develops computerized information linkages with outside suppliers, purchasing documents disappear in an evolutionary manner.

Figure 5-1 illustrates buyer and seller communications in a fairly sophisticated EDI system. The elimination of paper documents requires a well-planned sequence of events. Before discussing the elimination of external documents, several important points should be highlighted about Figure 5-1.

First, even though the EDI system is electronically sophisticated, some purchase orders, acknowledgements, requests for quote, etc., are still sent by mail. For example, special orders for capital equipment may continue to be processed in a non-EDI fashion. Second, even though two

FIGURE 5–1
EDI System

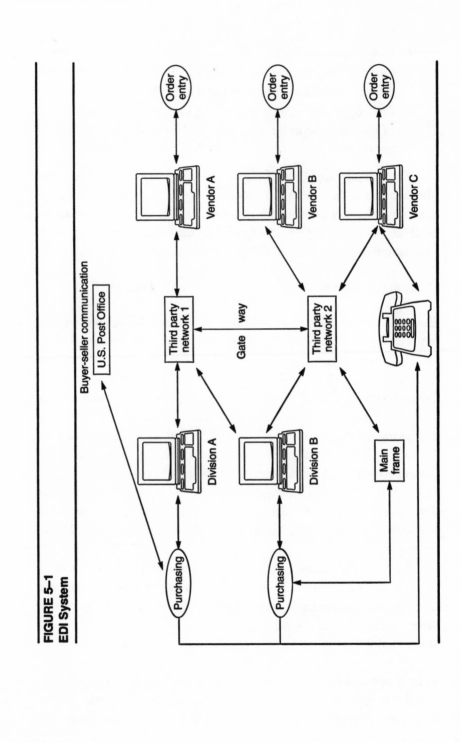

third-party networks are in operation, direct communication between the buying firm and one or more suppliers is available. This direct communication can be by computer, telephone, and/or mail. Finally, multiple third-party networks may be utilized via a gateway[2]. It seems probable that the use of third-party networks will increase significantly in the future. This suggests that many buying firms will use multiple third-party networks simultaneously, and these multiple third-party networks will need a gateway for cross-communication.

The first stage in the evolution to paperless purchasing is the development of EDI computer-to-computer linkages between a purchasing firm and a number of suppliers. The documents commonly eliminated in this stage are purchase orders, change order notifications, and purchase order acknowledgements. A major benefit of eliminating these documents is not simply paper reduction and clerical time saving, but, more importantly, reductions in the purchasing-order cycle time. This results in a shorter order lead-time for the buying firm's customers, a decrease in the inventory investment, and increased competitive advantage.

As firms advance toward the second stage of evolution in EDI sophistication, more documents are added to those already being transmitted electronically, the most notable being the request for quotation (RFQ) and the supplier response to the RFQ. More suppliers are also added to those communicating with the buying firm through EDI.

A very important linkage occurs during the third stage of EDI evolution. It is between the EDI system and the existing manufacturing planning and control system. This linkage is critical as a firm endeavors to cut the order cycle time further and, at the same time, increase the flexibility of the purchasing and manufacturing control systems. For example, one firm stated that EDI resulted in electronic communication with the internal customers of their own manufacturing facility (e.g., other plants, field distribution facilities, depots, and refurbishing centers). This resulted in significant savings in indirect labor and reductions in paper usage, lead times, and inventories.

The fourth stage of EDI development begins to "close the loop" in the electronic procurement system through the use of supplier bar codes. Computerized links can be developed with receiving and accounting. The linkage with receiving sets the stage for a reduction in paper documents, an increase in inventory accuracy, and a more timely exchange of shipping data. Some firms have questioned the need for any physical order shipment documentation. The linkage with accounting allows the electronic transmittal of invoices; the electronic comparison of invoices, receiving reports, and purchase orders; and the payment of invoices through electronic funds transfer. For example, the author has discovered that EDI linkages to accounting in conjunction with the bar coding of supplier shipments had a positive impact on firm operations. This linkage resulted in a quicker payment cycle, implementation of summary bills, and a reduced need for a paper verification system. Also, the supplier's use of bar codes to generate advanced shipping notices led to the implementation of planned receipts and shipments to point-of-use. Stage four is where the implementation of supplier bar codes has the greatest positive potential. The EDI linkages with receiving and accounting cannot fully develop unless the EDI loop is closed through the use of the bar code technology.

THE BAR CODING INTERFACE

What are the characteristics of an integrated EDI and supplier bar code system? If the systems are well planned, those characteristics are:

- Enhanced personnel performance monitoring.
- Reduced equipment costs.
- New hire avoidance.
- The balancing of incoming material receipts.
- The efficient use of existing equipment.
- The elimination of paperwork.
- Increased inventory accuracy.
- Increased material throughput speed.
- Improved shipment scheduling.

Integral to the EDI/bar coding process is a need to integrate (1) the EDI system, (2) the automatic identification (bar coding) system, and (3) the receiving and material handling systems. Each contributes an important part to the automation process. The integration of these three related functions is the key to unlocking the maximum productivity potential of the EDI/bar code system. Only when a system couples real-time control of material receipts with the data processing and communication capabilities of an EDI system is the procurement cycle fully automated and integrated.

The fully integrated purchasing information system will provide several vital pieces of information in real time. For example, data from the vendor's open purchase order file can be directly viewed and utilized by receiving. There can be automatic reconciliation of purchase orders with receiving reports, automatic distribution of products to point-of-use locations, and reserve storage and shipping. Also, there can be an automatic inventory update of material on hand.

The Bar Code Information Cycle

The EDI/bar code information cycle begins when a purchase order is electronically transmitted to the supplier. Against this purchase order, daily material order releases will be transmitted to the supplier automatically using the EDI network. At the supplier site, the order release, received through the third-party network mailbox, places the supplier bar code system into action. The information contained in the electronic order release is used by the supplier to create a bar code shipping label. At the time of shipment, this label is scanned by the supplier to generate an electronic advanced shipping notice (ASN) and a physical packing slip. The ASN is transmitted by the supplier, over the EDI network, to the buying firm just before the shipment leaves the supplier's loading dock.

When the shipment arrives at the buying firm's receiving station, the bar code label is scanned immediately. This information is automatically cross-referenced for accuracy with the ASN, order release, and purchase order. If a mismatch occurs, purchasing is informed and resolves the discrepancy. This same scanned information is used by the manufacturing planning and control system to update inventory information and release production orders. This process is further explained in the following example.

Honda of America Manufacturing., Inc.[3]

Honda of America Manufacturing., Inc. was one of the first firms in the United States to combine the EDI technology with bar coding in their

East Liberty, Ohio, complex. In the initial stages of the implementation, Honda quickly discovered that very few of their suppliers had experience matching EDI and bar codes. It was obvious to Honda that communication, education, and training of suppliers would be a critical element in the EDI/bar code implementation.

The purchasing information flow begins at Honda with Honda's monthly purchase orders. Against these purchase orders, Honda releases daily orders giving the supplier authorization to ship a specific amount, at a specific time, on a specific truck. This information is transmitted by Honda electronically using the EDI system to a third-party network mailbox where the supplier can electronically pick up the information at a specified time.

This transmitted order release information is then used by the supplier to create a bar code label to be attached to the shipment. This label is scanned by the supplier at the moment of shipment to create an advanced shipping notice (ASN) which is transmitted electronically using the same EDI network path back to Honda. At the same time, the supplier uses the scanned bar code information to create a physical packing list.

When the material arrives at the Honda loading dock, the packing list is scanned and compared with the material release for accuracy. Purchasing is notified immediately of any discrepancies. The scanned information is automatically made available to the production planning and inventory control system. This real-time, accurate information is critical to Honda's JIT manufacturing system.

Some Implications

The implementation of EDI and supplier bar codes can create a more effective receiving function through the better scheduling of receipts. As the paper disappears, the receiving function is forced into a closer integration with the ordering function. The information concerning order status and shipping, which at one time was available primarily to purchasing, can now be used by other functional areas for a more efficient operation. With the advent of more frequent orders and a shorter purchasing cycle time, this enhanced receiving operation is important to the proper functioning of the procurement and manufacturing systems.

The preceding discussion has several implications for managers. It is obvious that implementation of EDI, bar coding, more advanced procurement systems, and movement toward paperless purchasing will reduce costs. However, the benefits of paperless purchasing go far beyond just monetary savings.

The removal of paper documentation from the ordering and receiving systems compels the purchasing function into a fuller integration with other areas such as accounting, quality assurance, receiving, and manufacturing. The benefit of such an integration lies in the flexibility provided to the entire system. Such a company will be better able to exploit technological and market-based opportunities than its competitors.

THE COST/BENEFIT ANALYSIS

One critical task in implementing supplier bar codes is developing a cost/benefit analysis. Even though most of the participating firms indicated that a detailed cost/benefit analysis was not generally performed, it will be increasingly important to do so as more firms begin to implement the bar code technology with their suppliers.

Most technical advancements, such as bar codes, start with an individual's idea of how this technology can benefit the organization. Someone knows that a problem or opportunity exists and that the bar code technology may help. Fortunately, such intuitive analysis is not generally acceptable in a business environment. Business entities usually require a quantifying of these intuitive needs. A detailed cost/benefit analysis that serves to document needs and opportunities so they can be fully explored and exploited is appropriate.

To prepare such an analysis, certain data are required. Table 5-1 identifies the cost and benefit categories that are included in a *pre/post* bar code implementation analysis. Various costs are one-time developmental costs, while others are continuing. Table 5-2 shows how the specific costs associated with each category could be calculated. The exact figures associated with each category will vary from company to company. Each cost element should be included in a *pre/post* supplier bar coding analysis. Table 5-3 provides a sample *pre/post* analysis.

The system requirements should be clearly stated, along with their fit into the information and material flow, their financial justification, and the procurement and implementation options. Note that a complete understanding of the system's requirements is needed before a financial justification for them can be undertaken. The exact type and degree of support information necessary for the new system will depend upon the areas that were selected for improvement. Certain fundamental data should be obtained early in the project's life cycle.

ORGANIZING FOR SUPPLIER BAR CODE IMPLEMENTATION

Organization of the supplier bar code effort plays a critical role in the implementation process. Numerous alternatives exist concerning how a firm might organize for this implementation project. Two main principles should be followed: (1) clear leadership for the effort must be established, and (2) the effort should be multifunctional in scope.

In a large firm, representation from each functional area may be desirable. In a smaller firm, just a few people with cross-functional knowledge may be sufficient. It is the role of this implementation team to examine the project scope and to select the initial and subsequent pilot programs, to oversee the development of these programs both within and outside the firm, and to transfer technology and lessons learned from supplier to supplier. The supplier bar code implementation team will consist of three types of individuals: (1) technical people, (2) operations people, and (3) purchasing people.

TABLE 5–1
Cost/Benefit Analysis

Cost/Benefit Category	Cost Movement Direction
Managerial direct salary and fringe benefits	Down
Clerical direct salary and fringe benefits	Down
Paper document costs	Down
Document handling	Down
Telephone expense	Down
File space	Down
File cabinets	Down
Document storage	Down
Material handling system	Down
Data error costs	Down
Inventory costs	Down
Purchase price /cost	Down
Project team costs (direct & indirect)	Up
Hardware and software	Up
MIS support	Up
Materials and supplies	Up
Maintenance costs	Up

TABLE 5–2
Cost Category Determination

Cost Category	Calculation
Personnel cost associated with supplier contracts due to transaction document information problems	Average time per call multiplied by the salary rate multiplied by the annual number of transactions
Cost of paper transaction documents	Number of transaction documents per year multiplied by the unit cost per document
Paper handling	Average time spent handling a document multiplied by the annual number of documents multiplied by the clerical salary
Telephone expense associated with supplier contact	Average cost of a call multiplied by the annual number of required calls for problem resolution
Floor space costs for storage of transaction documents	Floor space for files multiplied by average cost of floor space
File costs	Number of files required multiplied by cost per file
Storage requirement costs for specific transaction documents	Floor space costs plus file costs for long-term storage of old files
Maintenance charges	Year cost of supplies and services for maintaining bar code equipment
Labor filing expense for transaction documents	Average time per filing multiplied by the clerical rate multiplied by the annual number of transactions
Software	Fee for the bar code and/or interface software
Data error costs	Average number of document and input errors multiplied by the cost to correct those errors
Personnel hiring avoidance	Number of new hires avoided multiplied by cost per hire
Inventory costs	Dollar value of inventory change multiplied by annual inventory carrying costs

TABLE 5–3
Cost/Benefit Analysis Worksheet

Costs	One-time	Recurring	Comments
Managerial direct salary and fringe benefits			
Clerical direct salary and fringe benefits			
Hardware charges			
Software charges			
Supplies and maintenance			
MIS support			
Project team costs			
Total costs			

Benefits	Current	Percent Improvement	Annual Dollar Benefits	Comments
Personnel cost associated with supplier contacts				
Paper transaction documents				
Paper handling				
Telephone expense associated with problem resolution				
File cabinets and space costs				
Long-term storage requirements				
Clerical expense				
Inventory costs				
Purchase price/cost				
New hire avoidance				
Total benefits				

Figure 5-2 illustrates an organizational structure that could be established for the supplier bar code effort. The steering function should be performed by a committee consisting of higher-level personnel from purchasing, operations, systems, etc. The mission of this committee is to provide directional guidance, resources, and schedules to further the supplier bar code implementation effort. Some firms have decided to give the steering committee a larger purpose. In these firms, the steering committee is assigned the task of directing the bar code efforts and applications in multifunctional areas.

Regardless of the size or composition of the team, the implementation team must have a leader, (i.e., a project director/coordinator). The leader should understand the company functions, both internally and externally. The team leader should also have a working knowledge of company MIS systems as they impact the receiving, purchasing, EDI, and materials handling systems. The leader will undoubtedly become the internal source of information on the bar code technology. This leader must have the ability and authority to define and implement the supplier bar code system. Finally, the team leader will act as a salesperson to sell the benefits and requirements of the implementation to superiors, subordinates, and peers.

The systems and technical development committee would primarily be concerned with developing all of the technical aspects associated with the supplier bar code effort. For example, this committee would consider using an established industry symbology and format standard, if applicable. It would also be responsible for evaluating the bar code software and hardware and how the bar code subsystem would interface with the existing MIS system.

The implementation committee should be charged with all tasks associated with introducing bar codes into the workplace. This would include all education and training (internal and external), auditing procedures, and any other changes in policy, procedure, and practice. It is imperative that the coordination and communication be frequent and clear between the steering, systems and technical, and implementation committees. Therefore, one or two individuals should have joint membership on all three committees.

FIGURE 5-2
Supplier Bar Codes—Organizational Alternative

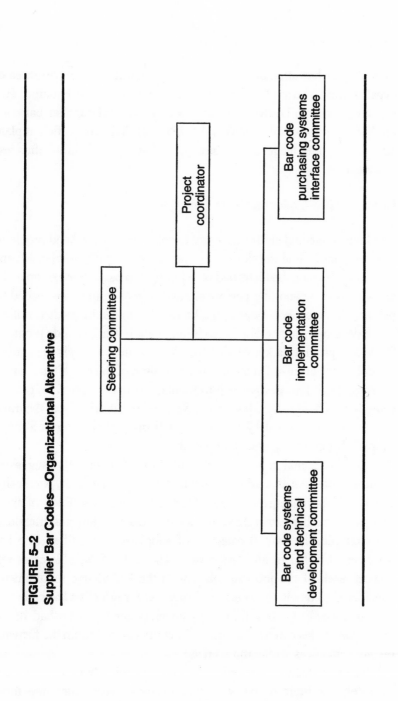

Steering committee

Project coordinator

Bar code systems and technical development committee

Bar code implementation committee

Bar code purchasing systems interface committee

LESSONS FOR MANUFACTURING

This chapter has attempted to distill and synthesize the experiences of over 20 firms who are leaders in supplier bar code implementation. Two factors seemed to be the cornerstones of successful supplier bar code implementation—*PLANNING* and *SUPPORT*. Without sufficient planning and support, the results of the supplier bar code efforts are often less than desired.

Impact of Bar Codes on Other Functions

The knowledge and skills necessary to solve bar code-related problems are job specific, and the skills necessary to successfully implement and use bar codes must be identified on a job-by-job basis. For example, the skills needed by auditing personnel are different from those needed by purchasing personnel. Developing knowledge and skill needs is critical to the development and implementation of a relevant bar code program. Table 5-4 provides a knowledge and skills content analysis for various bar code user groups. On the left-hand column of the exhibit is a listing of suggested knowledge requirements, grouped into five general categories. These categories were developed through study of the training manuals submitted by leading-edge firms participating in a major supplier bar code implementation study.

The horizontal row at the top of Table 5-4 lists the suggested training audiences divided into three major groups: supplier, purchasing and materials management, and interfacing functions. Each of these groups are further categorized into a list of specific job classifications. For example, the overall category of suppliers is classified into key managers, MIS/users, and salespeople. The body of the table cross-references each of the job categories with the knowledge requirements considered to be either necessary or desired for effective bar code task performance. From this figure, an understanding can be had of the impact bar codes can have on other functional areas within the firm and on work practices within those areas.

The introduction of supplier bar codes benefits both the buyer and supplier. As their delivery and receiving systems become more

TABLE 5–4
Knowledge Requirements

General Purchasing and Supplier Issues	Supplier			Purchasing and materials management			Interface functions			
	Key Managers	Salespeople	MIS/users	Management	Buyers	Receiving	diting	MIS	Accounting	Legal
1. Purchasing in the bar code environment	X	X		X	X	O	O	X	O	O
2. Organizational changes with bar coding		X		X	X	O	O	O	O	O
3. Task responsibilities with bar coding	O	X	X	X	X	X	X	X	O	O
Bar Coding Systems										
4. General overview	X	X	X	X	X	X	X	X	X	X
5. Paperless receiving	X	X	O	X	X	O				
6. Bar code to computer networks		O	X	O	O	O	O	X	O	
7. Costs and benefits	X	O	O	X	O		O	O	O	
Transaction Format and Content										
8. Role of standards	X	X	X	X	X	O	O	X	O	O
9. Industry standard used		X	O	O			O	X		
10. Label format and content		X	X	X	X			X		
11. Data segments and elements				O	X					X
Technical Issues										
12. Bar coding hardware		O	X				O	X		
13. Labels		X	X	O	X	O	O	X	O	
14. Translation software					X					X
15. Application software interface							O	X	O	
16. Symbology		O	O	X	O	O	O		X	
17. EDI interface	X	X	X	X	X	X	O	X	O	
Implementation Issues										
18. Implementation tasks			X	X	X	X	X	O		X
19. Responsibilities	X	X	X	X	X	X	X	X	O	O
20. Schedule			X	X	X	X	O		O	X
21. Policies, practices, and controls	X	X	X	X	X	X	X	X	X	X
22. Troubleshooting	O	X	X	X	X	O	O	X		

X = Necessary   O = Desirable

synchronized and interdependent, timely and accurate data are a necessity. Ultimately, the bar code system strengthens the buyer supplier relationship. Two important results ensue. Both parties become more flexible and competitive in the marketplace, and both experience operating cost reductions through the elimination of paperwork and the overhead associated with the paper flow process.

This chapter presupposed the need for a bar code system within receiving. The following example presents a counterargument.

> Another company, while evaluating a bar code system, recently discovered that its justification for the system disappeared when it eliminated the needless paperwork that had flowed among receiving, inspection, accounting, and production. The original projection had been for a two-year payback on the bar code system (based on the elimination of the clerical workers needed to produce the paperwork), but closer examination showed that most of that clerical reduction would come from just eliminating the unnecessary transactions.[4]

Given the conflicting opinion concerning the need for a bar code system, a transaction analysis should be the first step in an implementation effort.

ENDNOTES

1. It should be noted that most of these documents are eliminated with the implementation of a computerized purchasing and manufacturing control system. Clearly, significant productivity and competitive gains can be achieved through such systems.

2. In its most fundamental state, a third-party network resembles an old-fashioned pigeonhole mail rack. Each sender/receiver has a specific location which is identified and which has a security device to prevent unauthorized use. Third-party networks address the problems of different transmission modes, protocols, transmission speeds, time zones, and transmission loads between a buyer and supplier.

3. The information source for this example was the article "'Auto ID': Honda Marries EDI and Barcoding in New Plant," *Midwest Purchasing Management*, Murdoch Johnson, Editor, (March–April 1990), pp. 34–36.

4. Jeffrey G. Miller, and Thomas E. Vollmann "The Hidden Factory," *Harvard Business Review* 65 no. 5, (September–October 1985), pp. 142–150.

CHAPTER 6

INTERNATIONAL PURCHASING: IMPLICATIONS FOR OPERATIONS

INTRODUCTION

The importance of the linkage between the purchasing function and the other functions within a firm is steadily growing. Firms are becoming acutely aware that purchasing decisions impact many cross-functional decisions such as capacity requirements and equipment needs (make-versus-buy decisions), product cost and quality (source selection and qualification decisions), delivery reliability (mode and carrier decisions), and product innovation (supplier partnering decisions). This awareness is a result of the strategic emphasis being placed on product quality, cost competitiveness, and the just-in-time approach to manufacturing and delivery.

U.S. firms, recognizing these trends, are entering the arena of international purchasing in increasing numbers. Simply stated, domestic suppliers alone cannot meet all the competitive needs of a multinational corporation. As a result, international sourcing has emerged as a critical component of corporate strategy aimed at reducing costs, raising product quality, increasing manufacturing flexibility, and improving designs.

This chapter discusses the strategic importance of international purchasing, describes procedural and managerial issues of concern in international procurement, and delineates the differences between buying from foreign versus domestic sources.

A CHANGING BUSINESS ENVIRONMENT

There are several reasons for the emergence of international purchasing as a strategic weapon in the restructuring of manufacturing operations in U.S. firms. Almost all of these reasons are directly related to the efforts of U.S. manufacturing firms to gain (or regain) competitive strength and market share by improving their strategic posture in response to a changing business environment. The principal changes in the business environment that underlie the move by manufacturing firms to evolve new strategies are listed in Table 6-1.

Competition

Intense competition from abroad, pressures stemming from the need to reduce trade deficits through exports, and the interdependence of global economies have all served to internationalize the marketplace. Surviving and thriving in today's global markets require that manufacturing firms be truly "world class." There is evidence that the United States no longer enjoys an unequaled advantage manufacturing and in manufactured goods. Steel, semiconductor, automobile, and consumer electronics industries all have fallen victim to international competition and have seen their market shares erode.

Cost Reduction

Manufacturing firms are striving to be the lowest-cost, highest-quality producers. For firms selling in mature markets where there is little or no

---

TABLE 6–1
A Changing Business Environment

New Challenges To Corporate America

Intense international competition.
Pressure to reduce costs.
Need for manufacturing flexibility.
Shorter product development cycles.
Stringent quality standards.
Ever-changing technology.

---

product differentiation, cost reduction is especially important. The recent experience of U.S. industries in steel, consumer products, and automotive manufacturing underscores the importance of cost and quality as competitive weapons. Cost, quality, and customer satisfaction have become the foundation on which a successful competitive strategy is built. It is well known that the Pacific Basin countries enjoy a comparative advantage in cost. For example, the average labor costs in the electronics industry are shown in Table 6–2.

It has been suggested that lower wages abroad give foreign manufacturers a substantial cost advantage over U.S. manufacturers. In the case of automobiles, this cost difference has been estimated at between $1,500–$2,000 per vehicle. In light of this, domestic manufacturing firms have relentlessly pursued the reduction of labor costs. One way of achieving this objective was to reduce the labor content of manufactured products through extensive automation. Ford, General Motors, and Chrysler serve as prime examples of the firms spearheading manufacturing automation.

Another method used by domestic manufacturers to reduce labor costs is to purchase labor-intensive components from foreign suppliers who operate in cheaper labor markets. The motivation for this stems from the distinct cost differentials that exist between the U.S. and other

TABLE 6–2
Comparison of International Labor Rates

	Electronics Industry	
Country	*Fringe Rate**	*Dollars Per Hour*
Malaysia	25%	$1.44
Hong Kong	25	1.45
Taiwan	60	2.26
Singapore	35	2.59
South Korea	80	2.72
United States	46	11.90

*Percent of hourly rate.
Source:Machinery and Allied Products Institute., *MAPI Survey on Global Sourcing as a Corporate Strategy*, (Washington D.C., 1986)

countries, (e.g., the Pacific Basin). Honeywell, for example, buys electronic assemblies and components from Japan, Hong Kong, and Korea. Ford, GM, and Chrysler also spend a significant portion of their purchasing dollars offshore.

Of the three major domestic automobile manufacturers, Chrysler is the most active in international procurement operations. The number of in-house hours it takes Chrysler to build a car has dropped by 40 percent, a reflection of Chrysler's out-sourcing strategy. This strategy seems to be favored by companies which market mature products where advantages stemming from product differentiation are minimal. In these mature markets cost advantages are the key determinants of competitive edge.

Need for Manufacturing Flexibility

Investments in capital equipment are becoming ever more costly. Capital investments, however, are necessary to improve productivity and to compete effectively in international markets. The need to ensure manufacturing flexibility, which enables a firm to introduce new products more quickly and to parry competitive threats by modifying existing products, also necessitates automation and investment in capital equipment. Increasingly, U.S. manufacturing firms rely on suppliers to acquire new technology and manufacturing capacity.

Manufacturing firms are turning to suppliers for routine manufacturing of components and subassemblies. In the case of the electronics industry, more than 70 percent of the content of manufactured products is purchased from outside sources. Westinghouse uses outsourcing and offshore buying as corporate strategies to enhance manufacturing flexibility. As part of the strategy to ensure and enhance flexibility, firms pay particular attention to offshore suppliers with demonstrated engineering, technological, and process capabilities.

Product Development Times

Competitive pressure to reduce the cycle time from product design to market introduction continues to exert an important influence on manufacturing operations. For example, the cycle time in the automotive industry for domestic manufacturers is approximately four to five years, compared to an average of two years for Japanese manufacturers. This gives the Japanese the ability to exploit rapidly changing market opportunities before domestic automakers can respond.

In order to reduce development time, the concept of "simultaneous engineering" is gaining acceptance from domestic manufacturers. Simultaneous engineering involves the supplier firm early in the design stage of a new product development cycle. The benefits of better integration with a supplier's capabilities through better communication, higher quality and less time to develop a product accrue to the buying firm through a partnership with the supplier firms. The benefits of longer-term and more predictable contracts, the opportunity for a "strategic partnership," the sharing of technological expertise, assistance in developing process and product capabilities, and playing a role in the long-term viability of the buying firm accrue to suppliers. Today, American firms are searching the world for suppliers with the technical expertise to assist in reducing the buying firm's product development lead times.

Quality

Manufacturing's resurgence within the functional hierarchy of a firm is largely attributable to the realization that quality is essential to the very survival of a firm. Japanese firms have achieved unparalleled success and have become world leaders in manufacturing principally through their emphasis on quality. Product quality is now the undisputed, central issue in manufacturing. The conventional wisdom which held that there is a trade-off between quality and cost has been replaced by the belief that both high quality and low cost are attainable through diligent planning, careful design, and efficient and effective processes. The attainment of zero defects is the manufacturing goal. The dramatic improvement in quality achieved by some domestic manufacturers provide examples of this trend.

The impressive turnaround in quality achieved by Control Data Corporation at its disk drive plant in Minnesota typifies efforts of U.S.

firms on the move to regain competitiveness. In 1985, the cost of scrap alone at CDC's disk drive operations was running at $3 million a year, which was approximately 37 percent of the manufacturing budget of the plant. A total quality improvement effort was undertaken, and within two years, 95 percent of the disk drives produced by CDC were defect-free and required no rework. The dramatic improvement in quality has helped CDC stem market erosion to its Japanese competitors. Figure 6-1 shows the path toward zero defects taken by CDC.

Xerox's turnaround in the office products market has been achieved on a solid foundation of quality improvements. It can now build a copier for about half of what it would have cost to build a few years ago. In 1984, parts supplied to Xerox Corp. were roughly five times as likely to fail as parts supplied to Fuji Xerox Company Ltd. of Tokyo. In 1986, after intensive efforts to improve the process capabilities of its suppliers, Xerox achieved part quality parity with Fuji Xerox. Figure 6-2 shows Xerox's impressive gains in supplier quality.

These improvements in quality were achieved, in the main, without burdensome investments in automation. Total quality programs were the centerpieces of these remarkable achievements. Both these firms acknowledge that an increase in foreign sourcing was an important component of their total quality improvement efforts. According to Boston University's Manufacturing Futures Research Project, total quality programs have displaced automation approaches as the most frequently used action programs for effecting manufacturing change.

New Technology

Technological innovation is essential for maintaining and improving market share as a product evolves through its life cycle. It is well documented that U.S. industry has not been investing in and automating capital equipment at the same rate as foreign competitors. The slower pace of investment in capital and in research and development has hurt the ability of U.S. firms to introduce new products and product innovations at the same pace as their Japanese counterparts. The success of Japanese firms in such consumer electronics products as VCRs, compact disc players, and TVs is a case in point. Technological innovation is now

FIGURE 6–1
Control Data Learns to Control Defects

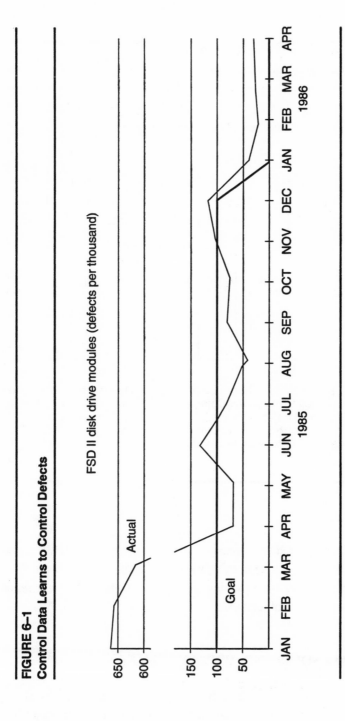

FSD II disk drive modules (defects per thousand)

Source: Norm Alster, "High-tech Manufacturing," *Electronic Business*, 1987, pp. 53–64.

FIGURE 6–2
Xerox Duplicates Japanese Quality

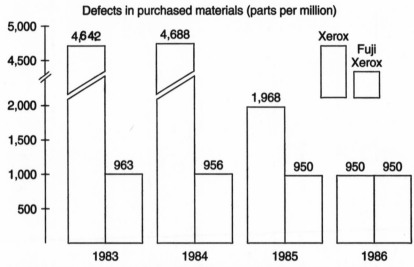

Defects in purchased materials (parts per million)

In 1984, parts supplied to Xerox Corp. were roughly five times as likely to fail as parts supplied to Fuji Xerox Company Ltd. of Tokyo (Xerox's joint venture with Fuji Photo Film Company Ltd.). In 1986, after intensive efforts at improving the process capabilities of its suppliers, the U.S. copier giant claims part quality parity with Fuji Xerox.

Source: Xerox Corporation.

recognized as both a defensive strategy against competition and an offensive strategy in aggressively seeking to build market share. Foreign sources of supply are now viewed as prime sources of new technology for American manufacturers.

GLOBAL SOURCING: WHY PURCHASE ABROAD?

Faced with these changes in the business environment, U.S. firms are turning to global sourcing as a competitive strategy. Data provided in

Table 6-3 attest to the fact that international purchasing is becoming pervasive in U.S. firms. The cost advantage enjoyed by the Pacific Basin countries provides an impetus for U.S. firms to seek qualified manufacturers in these countries as potential suppliers. In a recent symposium held at Michigan State University, Allen Meilinger, corporate vice president at Westinghouse Corporation, observed, "Gaining competitive advantage is our corporate goal. Global sourcing is a key strategy in accomplishing this goal." This strategic thinking exemplifies the emerging trend in sourcing practices in the United States.

The amount being spent by U.S. firms on offshore purchases is sizeable. It is estimated that nearly 10 percent of Chrysler's $8.6 billion is spent on global purchases; Honeywell's material expenditures on international purchases exceed $200 million; and that Westinghouse Corporation spends roughly 7 percent of its total purchases on international procurement. The amount of purchasing dollars being spent on offshore purchases has grown from an average of 11 percent a few years ago to 15.5 percent in 1986.

What are the principal reasons for this recent surge in offshore buying? Some of the answers to this question can be gleaned from the 1986 survey of 80 large U.S. manufacturing firms conducted by the Machinery and Allied Products Institute (MAPI) of Washington, D.C. The respondents to the MAPI survey ranked lower cost, improved quality, manufacturing flexibility, and access to new technology, in that order, as being the most important reasons. Although lower costs and improved quality are most frequently associated with offshore pur-

TABLE 6-3
Foreign Purchasing: A Growing Phenomenon

Annual Sales	Percent of Firms that Import
Under $1M	33%
$1–$10M	37
$10–$100M	54
Over $100M	65

Source: *MAPI Survey on Global Sourcing as a Corporate Strategy*, Machinery and Allied Products Institute, Washington, D.C., 1986

chases, the capability of the suppliers to provide technical support and increase manufacturing flexibility also plays an important role in companies' decisions to buy offshore. Several additional executives interviewed by the authors were nearly unanimous in stating that cost, quality, and technology were the primary reasons for offshore buying. The diversity of goods and services bought from foreign sources is reflected in Table 6-4, which shows the proportion of survey respondents who source abroad for each category of purchase. These proportions indicate that international purchases are broadly diversified among materials and manufactured goods, such as machinery and subassemblies.

Even though the purchased items are diversified, the location of purchase for these items were not. A follow-up study performed by the authors indicated that the majority (60 percent) of items sourced from the Pacific Basin were electronic components and subassemblies, while the majority of items sourced in Europe were of the machinery and equipment type.

THE PRACTICE OF INTERNATIONAL PURCHASING

The purchasing cycle for foreign sourcing, outlined in Figure 6-3, has several distinct phases: (1) recognition of need; (2) source identification; (3) source evaluation; (4) evaluation of quotations; (5) subjective analysis and negotiation; and (6) contract management. Even for a domestic buy, the purchasing task is complex. A purchasing department deals with hundreds of sources for thousands of items. This creates a

TABLE 6–4
Foreign Sourcing Practices: Items Purchased Abroad

	Percent of Foreign Purchases
Materials	76%
Machinery and equipment	69
Component parts	81
Services	16

difficult administrative job in addition to purchasing's main task of skillful buying. International purchasing involves a series of tasks similar to domestic buying, but these tasks differ in their complexity and level of detail. The principle differences occur in the activities described in the following sections.

Source and Product Identification

There are several ways in which firms interested in international purchasing can identify potential sources of supply. Sources of information such as commercial attaches, large banks, U.S. government documents, international trading companies, and the U.S. State Department are representative of those being used by manufacturing firms engaged in international sourcing.

At Chrysler Corporation, source identification is preceded by an organizational "needs analysis." This is done through a survey of organizational units by the office of the director of worldwide procurement. The purpose of the survey is to identify unmet needs pertaining to product, technology, quality, or some other service. Based on the needs assessment, a search is initiated to identify world class suppliers.

A number of considerations go into selecting products suitable for overseas sourcing. The factors that exert an influence on this selection are the long supply lines (distance of suppliers), the need to clearly communicate engineering specifications, terms and conditions of the purchase contract (Control Data, for example, translates its terms and conditions into Kanji for Japanese vendors), and the extent of supplier development required (including site visits). The following list of considerations used by Honeywell typifies criteria used in selecting products for overseas sourcing:

- Stability of design.
- Duration of anticipated association with vendor.
- Whether or not the product requires continuous runs.
- Completeness of engineering and other documentation.
- Ability to provide technical/quality assistance.
- Materials and tooling required.
- Necessary visits.

FIGURE 6–3
International Purchasing Cycle

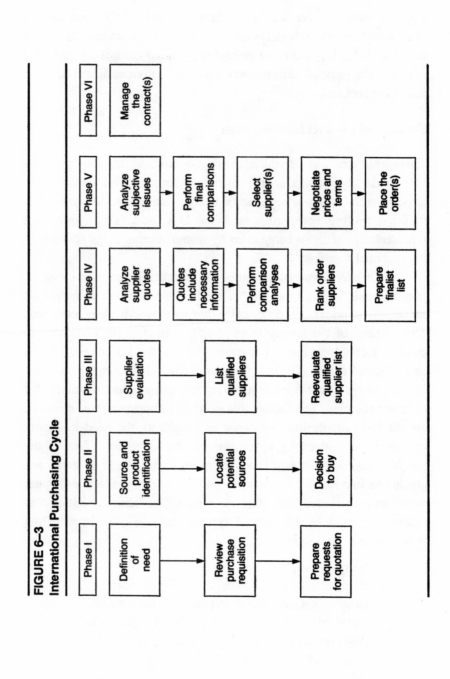

Qualifying Sources

Should a firm use a foreign source that is deemed not fully capable of performing? In the domestic market, a company has legal recourse against a supplier who fails to honor contractual agreements. This same legal recourse either may not exist or be too burdensome when dealing with a foreign supplier. A good rule of thumb is to reduce the risk of nonperformance by qualifying the foreign supplier before allowing the supplier to bid on a contract.

Considerations important in foreign source qualification are experience of the supplier as a foreign source and manufacturer, financial strength of the supplier, the ease with which effective communication can be established, and implications for inventories (size, location, and so on).

Experience—A majority of the respondents in a study done by the authors to identify principles and practices of international purchasing mentioned that a careful evaluation of the supplier's experience and management expertise is essential to selecting a reliable foreign supplier. It is advisable to ask for and check references from companies doing business with the foreign supplier.

Financial Strength—The financial strength of a supplier needs to be checked carefully. The company should be capable of meeting the increased expenditures necessary for equipment, marketing, and additional inventory. The director of corporate procurement for A.B. Dick Company in Chicago suggests $10 million in sales and 100 employees as the lower limit for the size of the foreign supplier.

Communication Lines—It is essential to ensure that good communication lines exist to foreign sources. U.S. firms could require a foreign supplier to designate a U.S. representative who can ensure that communication lines are kept open. As the involvement of a firm in global sourcing increases, the need for developing communication systems to support source identification, supplier development and qualification, relocation of purchased materials, and inventory controls will rapidly increase.

Inventory—Foreign sources must be willing and able to maintain higher levels of inventory to compensate for the longer supply lines, tight specifications, and stringent delivery requirements.

Longer-term orientation—Because of the length of time it takes to identify, develop, and qualify foreign sources, it is important for firms to strive for a long-term association with a supplier.

Analyzing Quotations from Foreign Sources

Requests for quotation (RFQ) issued by a manufacturing division are generally routed through an offshore procurement office, a trading agent, a broker, or the company's subsidiary. The offshore procurement office or the trading agent handles the transmittal of the RFQs to potential suppliers. This office generally performs such liaison functions as vendor searches and surveys, distribution of RFQs, transmittal of proposals from vendors, and so forth. At Honeywell Corporation, field offices are generally staffed by engineers with considerable marketing experience. In the Far East, these offices are staffed with foreign nationals drawn from the region.

The proposals from suppliers in response to RFQs are sent back to the manufacturing division for evaluation. Some cost elements that should be considered in comparing proposals from foreign suppliers are included in Table 6–5. Purchase orders are released by the individual divisions to the suppliers through the corporate international procurement office. These general practices are used, for example, by Chrysler, Control Data, and Eaton corporations.

One of the principal functions performed by the corporate international procurement personnel is ensuring that foreign sources clearly understand the procedures needed to bid on a contract. They also have the responsibility to assist in interpreting the engineering and manufacturing specifications and answering questions pertaining to them. Typical quotations ask the suppliers to provide information on:

- Annual volume and order quantities.
- Unit costs for quantities shown in U.S. dollars.
- Tooling costs.
- Cost of providing qualification samples.
- Manufacturing and delivery lead times.

TABLE 6–5
Foreign Sourcing: Cost Elements to Evaluate

1. Unit price
2. Export taxes
3. International transportation costs
4. Insurance and tariffs
5. Brokerage costs
6. Letter of credit
7. Cost of money
8. Inland (domestic and foreign) freight cost
9. Risk of obsolescence
10. Cost of rejects
11. Damage in transit
12. Inventory holding costs
13. Technical support
14. Employee travel costs

In addition, information is requested on the business practices of the vendor, including whether or not a letter of credit is required, the name of the bank handling the letter of credit, payment terms for open account transactions, principal customers of the foreign supplier, FOB point, and size and scope of operations (number of employees, annual sales volume, market share, etc.). Control Data, for example, recommends to its buyers that samples be exchanged with Japanese suppliers since the Japanese pay particular attention to workmanship and defects that can be detected visually.

The responsibility of the manufacturing divisions is to develop requirements specifications, specify vendor qualification criteria (done by divisional purchasing people), perform analysis, provide negotiation assistance, place orders, and cooperate with the corporate international procurement office. Shipment details, international carrier control, and dealing with customs are handled by physical distribution and traffic management personnel. Criteria for vendor qualification are often the same as for domestic suppliers. This was true in all the companies participating in this study.

Negotiating Prices and Terms

Negotiating prices and terms with a foreign supplier poses additional challenges to a buying firm. Cultural and language barriers exacerbate the difficult task of negotiating with a supplier. Negotiating might often require the services of an interpreter. Extensive preparation is required before serious negotiations can be undertaken with a foreign source. This preparation includes a study of costs, supplier's management strengths, supplier's growth potential, currency exchange rates, handling of rejected materials, and so on.

Increasingly, organizations are turning to consulting firms for cost analyses and projections for currency exchange rate projections. Arthur D. Little, of Cambridge, Massachusetts, makes available to its client firms a computerized model that projects the cost of manufacturing a specific component in Japan by analyzing its material content, its part geometry, and the manufacturing steps required to produce the component. The model utilizes data on labor, materials, energy, transportation, and factory overhead costs. Arthur D. Little's database can currently project costs for 80 products in 16 countries. The computer model equips a buying firm with detailed knowledge of a foreign supplier's present and future components of cost.

Besides costs, a checklist of purchase conditions might include specific carriers and modes of transportation from the foreign source to the United States, payment of transportation insurance, provisions for returning and/or replacing defective goods, and the method of payment. Payment arrangements are usually made by the international purchasing department, especially when an international letter of credit is involved.

Managing the Contract

Speakers at a seminar on overseas buying held at the University of Wisconsin-Madison recommended that buyers seek the assistance of a customhouse broker who can provide information on commodity class descriptions that permit the most favorable duties, special tariffs, and the effects (if any) of FDA regulations and regional differences in the interpretation and/or application of customs rules and regulations. Occasionally savings in customs duties can be realized simply by changing the point of entry into the United States, since the valuation of imported goods for duty purposes is often not consistent from one customs point to another. The levying of duties is also affected by international politics and the status of exporting nations as trading partners of the United States. Control Data Corporation clears all imports for each ordering facility at one prespecified customs point close to the ordering facility. CDC uses the services of a customs broker for clearing imported goods.

LESSONS FOR MANUFACTURING

Implementing an international purchasing strategy can be confusing at first, and the process is sufficiently delicate that it must be managed well to be successful. Figure 6–4 illustrates a flow of activities that a multinational company might use to establish an international purchasing practice. First, a company must formally plan for global purchasing. Multinational firms should establish a formal decision-making process by starting with a comprehensive analysis of needs before embarking upon international purchasing. This needs analysis process should identify procurement requirements, strategies for fulfilling these requirements, and their relationship to overall corporate and competitive goals. The needs analysis process should lead to the identification of commodities and components suitable for international sourcing. An assessment of material needs is best done under the direction of a corporate purchasing or materials management staff using a program-planning and budgeting (i.e., a bottom-up) approach. Such an approach forecasts material needs at the plant and divisional levels and aggregates up to the corporate level. This approach offers the advantages of comprehensiveness of planning and a fuller participation in the planning process by various organizational units.

Second, it is necessary to create an international procurement staff function at the corporate level responsible for gathering information (e.g., potential suppliers, price trends, etc.) and evaluating opportunities for buying abroad. Since international purchasing involves long supply lines, pipeline inventories, difficult negotiation, cultural differences, and communication problems, expertise in domestic purchasing does not readily translate to international sourcing. For example, the corporate staff must become familiar with the electronic transfer of information

FIGURE 6–4
Implementing International Purchasing

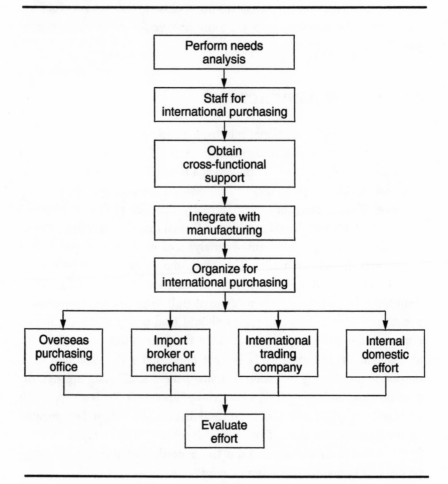

abroad, customs regulations, foreign modes of transportation, and the effect of exchange rate fluctuations on the cost of imported merchandise.

Third, a firm embarking upon international purchasing should obtain cross-functional support for this effort. Because of the length of transportation supply lines associated with purchasing abroad and the resulting increased potential for supply disruptions, a firm engaged in international purchasing needs to achieve a higher level of interaction with and support from other functional areas such as manufacturing, design engineering, and quality assurance. Achieving such a high level of cross-functional support is a challenge to most firms. Based on our interviews, it appears that the following activities might prove useful.

- Soliciting the direct involvement of top management.
- Formally integrating international purchasing within the corporate strategy.
- Designing organizational structures that support the cross-functional planning approach (e.g., matrix or product line structures).
- Emphasizing the value-chain approach to materials management by adopting a systems view of the procurement, manufacturing, and transportation functions.

Fourth, both the manufacturing and purchasing planning and control activities need to be precisely integrated. Such integration is critical to the success of international purchasing. For example, firms engaged in just-in-time deliveries and manufacturing need to fine-tune their inbound logistics systems to use international sources while keeping pipeline inventory levels at a minimum. Careless or inadequate planning will exacerbate problems, create excessive inventory and/or out-of-stock positions, destabilize production plans, and ruin customer service levels.

Fifth, firms engaging in international purchasing have a variety of alternatives available for organizing personnel. Which alternative is chosen depends upon such factors as the level of specialized foreign buying knowledge available within the firm and the expected volume and frequency of foreign purchases. Some of those alternatives are (1) an overseas purchasing office, (2) import broker or merchant, (3) international trading company, and (4) assignment within the corporate purchasing function. Each of these alternatives has advantages and disadvantages. Our research showed that most firms prefer overseas purchasing offices. These offices were primarily staffed by foreign nationals with technical backgrounds who were trained in purchasing.

Finally, the corporate procurement planning group should be continually evaluating the strategy and practice of buying from abroad, given changing business circumstances. Each firm should have a portfolio of suppliers—some domestic, others foreign. The allocation of purchasing dollars among this portfolio should be in accordance with corporate strategies aimed at cost reduction, technological leadership, quality competitiveness, and others. Careful consideration should be given to the length of purchasing contracts and the degree of integration needed with the supplier's manufacturing system.

Impact of International Purchasing on Other Functions

Internationalization of the marketplace, global competition, and changes in the business environment have contributed to the increase in international purchasing. The push to buy internationally is on, in both small and large U.S. manufacturing firms. Global sourcing will continue to grow as a matter of corporate policy. Companies interviewed for a study conducted by the authors all have corporate purchasing offices for global buying established under corporate policies. Between 1984 and 1987 the proportion of companies with corporate international procurement operations increased from 46 percent to 61 percent.

International purchasing opportunities are changing the mix of manufacturing firms (through make/buy decisions) and thereby affecting capital investment requirements and the manufacturing infrastructure within companies. For example, international transportation is increasingly becoming the responsibility of the purchasing function.

International purchasing has had a significant effect on manufacturing operations. Reductions in the number of components and manufactured subassemblies, implementation of JIT, an increase in the quality of manufactured goods, changes in manufacturing infrastructure, and closer cooperation among manufacturing, marketing, and purchasing personnel are direct consequences of international purchasing. International purchasing is a competitive weapon. Recognition of this has led to its growth in U.S. firms, a trend that is likely to continue. Manufacturing professionals will be called upon to play a different role as companies push toward integration of foreign suppliers into their manufacturing systems. Developing stable production plans, simpler designs, process designs that ensure a smooth production flow, and tighter linkages with purchasing and marketing will be more essential for manufacturing than ever before.

Some unique issues arise when the decision to buy abroad is made within the firm. Cross-functional input into the procurement process is mandatory, if potential problem areas are to be avoided. A key to effective international sourcing is selecting flexible suppliers of the highest quality. Suppliers' quality is often difficult to assess because data concerning suppliers and their respective quality performances are often not readily available. Firms sourcing overseas will make frequent visits to potential sources to assess supplier capability. Since these visits may require a team assessment, resource allocation constraints become paramount.

Most overseas products will be shipped by ocean, and the lead time will be several weeks. This means that the buying firm must plan capacity and material utilizations on a much longer time frame. Schedules must be stabilized or inventories can mount to unacceptable levels. The concept of just-in-time really does not exist when dealing with a foreign supplier.

The act of expediting a foreign supplier's shipments takes on a new meaning. The purchasing function must be on good terms with the foreign supplier's personnel. In fact, the quality of a foreign supplier's management is probably as important as the quality of the purchased product during the selection decision.

During the last 20 years, foreign currency exchange rates have floated freely with respect to the dollar. Fluctuations have been, at times, precipitous. This means that before signing a contract, purchasing must work closely with finance to forecast likely exchange rate movement scenarios and likely methods for moderating the impact for such fluctuations. Arbitrarily contracting for payment in U.S. dollars makes little economic sense. As the European Common Market moves toward a universal currency, purchasing must acquire expertise in areas that are relatively new. Chapter 8 of this book examines some of these issues in detail.

Quality problems from an overseas supplier can have an onerous impact on production. The pipelines are long, distances are great, languages are different, and misunderstandings common. Quality specifications need be understood by all parties before production is begun by the supplier. Engineering, quality assurance, and production must work together to assure the highest level of supplier quality. When rejects do occur, return, reimbursement, and replacement of the offending items can be quite complex. The bottom line: . . avoid defects at all costs.

CHAPTER 7

INTERNATIONAL COUNTERTRADE: LINKING PURCHASES TO MARKETS

INTRODUCTION

In simple terms, *countertrade* is a word used to include all international and domestic trade—indeed all trade—in which goods are exchanged for goods. The ancient business of trading a few sheep for a head of cattle has grown over the last 20 years into a sophisticated worldwide trading activity with significant economic impact on the worldwide market. Figure 7-1 shows how countertrade has grown from an estimated 2 percent of world trade in 1975 to a conservatively estimated 8 percent in 1985 and 10 percent in 1986. This 10 percent figure translates into over $150 billion in world trade in today's international economy. Likewise, Figure 7-2 graphically demonstrates that countries which mandate some form of countertrade by law have grown from a mere 15 in 1975 to over 88 in 1984.

In a 1985 survey of major U.S. multinational companies, the United States Association of Countertrading Corporations found that 85 percent of respondents had received countertrade demands from a total of 65 different countries. That same year, the U.S. International Trade Commission surveyed over 500 major U.S. companies and found that $7.1 billion of American exports in 1984 were linked to countertrade deals, and that armaments accounted for more than 80 percent of the countertrade deals. Exports associated with nonmilitary countertrade quadrupled from $285 million in 1980 to $1.4 billion in 1984. In both categories, the United States' main partners for countertrade were other industrialized nations.

FIGURE 7–1
Countertrade

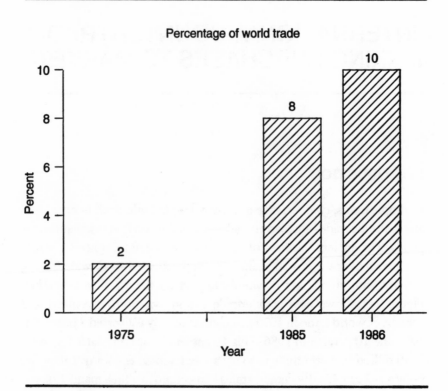

Percentage of world trade

FORMS OF COUNTERTRADE

Starting with simple trading of goods and services for other goods and services, the modern concept of countertrade has evolved into a diverse set of activities that include five distinct types of trading arrangements.

Barter is the direct exchange of goods or services or both between two parties without a cash transaction. Barter, although in theory the simplest trading arrangement, has become an unpopular countertrade arrangement. For example, if goods are not exchanged simultaneously, one party ends up financing the other for a period of time. Barter also risks sticking one of the parties with goods it does not want and cannot

---

FIGURE 7–2
Mandatory Countertrade

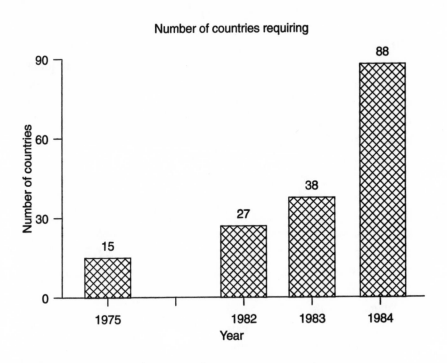

Number of countries requiring

use. For these reasons and others, barter is viewed by multinationals as the most restrictive countertrade arrangement and is primarily used for one-time deals with the least creditworthy or trustworthy trading partners.

Counterpurchase is reciprocal buying in the international marketplace. It occurs when a company agrees to buy a certain amount of materials from a country to which a sale is made. For example, United States company X sells some of its products to foreign country Y. In exchange for the sale, United States company X agrees to spend some set percentage of the monetary proceeds of the sale importing goods produced by foreign country Y. Both trading partners agree to pay for the majority of their purchases in cash and fulfill their individual sales obligations to each other over a specified time frame, usually less than five years.

Offset is similar to counterpurchase in that the supplier is required to accept an obligation to purchase goods or services up to an agreed percentage of the original sale. The difference is that the supplier can fulfill the countertrade obligation with any company in the country to which the sale is being made. Offset arrangements are most frequently performed with centrally planned economies. For example, in 1982 McDonnell Douglas Corporation agreed to buy airframe components and other goods from Canadian companies in exchange for a $2.4 billion commitment by Canada to buy jet fighter planes over the next 15 years.

Compensation or *Buy-Back* occurs when a company builds a plant in or supplies technology, equipment, training, or other services to a foreign country and agrees to take a certain percentage of the plant's output as partial payment for the contract. For example, Occidental Petroleum negotiated a $20 million deal with the U.S.S.R. in which Occidental would build several plants there and receive ammonia over a 20-year period as partial payment.

Switch Trading is the use of a specialized, third-party trading house in a countertrade arrangement. The third-party trading house buys the selling company's counterpurchase credits and sells them to another company who can better use them. For example, Company X concludes a countertrade arrangement with some country in which it agrees to take some amount of goods as partial payment. Company X cannot really use and does not want these goods. Therefore, Company X sells to a third-party trading house at a discount the contract for these goods. The third-party trading house in turn finds another company who can use these goods and resells the contract at a profit. These arrangements frequently substitute trade credits, spendable later, for cash. Oftentimes, switch trading is used by companies or countries to correct various types of trade imbalances. For example, if Hungary sells chemicals to France at a certain value, France credits Hungary's trade account for a certain amount of trade credits which Hungary can use in the future to buy some pre-specified French goods. But, Hungary doesn't want to buy French goods. Therefore,

Hungary sells these credits to a third-party trading house at a discount. The third-party trading house then searches for a country or company who needs to buy French goods and sells the credits to this country or company at a small profit.

In 1983, the National Foreign Trade Council (NFTC) conducted a survey of 110 U.S. firms that were active in the international market. Figure 7-3 presents the results of this survey. The NFTC found that 55 percent were engaged in counterpurchase, 24 percent were engaged in offset arrangements, 9 percent practiced compensation, 8 percent used switch trading, and only 4 percent were involved in barter agreements.

In a recent survey of Fortune 500 companies, a somewhat different mix of countertrading activities was found by the author. Figure 7-4 presents a graphic representation of the results of this survey. Offset was

---

FIGURE 7-3
Countertrade Practices, 1983

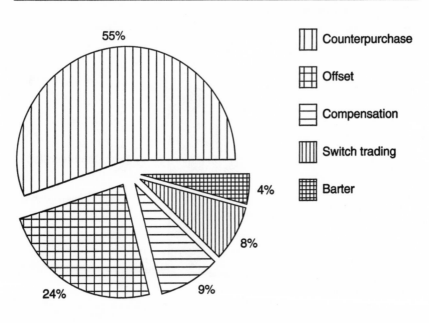

the most common type of countertrade arrangement, being practiced by 73 percent of the respondents. The next most common arrangement was counterpurchase, followed by compensation, barter, and switch trading being practiced by 60 percent, 22 percent, 19 percent, and 3 percent of the respondents, respectively. Several trends seem apparent. First, the survey results suggest that countertrade activities have increased during the 1980s. This is consistent with the previously mentioned National Foreign Trade Council study which showed that approximately 88 countries had some form of mandatory countertrade requirement in 1984. Second, companies are obviously practicing more than one form of countertrade simultaneously, with offsets and counterpurchases the most prevalent. Third, the much higher percentage of companies involved in barter, along with the drop in the percentage involved in switch trading, seems to be an indication that companies are getting directly involved in countertrade activities rather than relying on a third-party trading house to work out arrangements for them.

FIGURE 7–4
Countertrade Practices, 1986

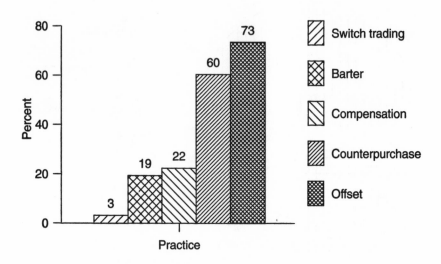

A COUNTERTRADE IMPLEMENTATION MODEL

Implementation of a countertrade strategy could be very confusing for the multinational company embarking upon countertrade for the first time. Furthermore, a complicated process such as countertrade must be well managed in order to be successful. An understanding of the options available to and actions required by the multinational corporation is required as it implements countertrade arrangements.

Figure 7–5 illustrates a process the multinational company might use to set up a countertrade activity. First, it should analyze the countertrade opportunity—see whether there is a countertrade opportunity within its operating environment. Identification of the opportunity or the perceived benefits may be driven by top-level executives, materials managers, or marketing and sales personnel. If the countertrade opportunity is perceived to be low, then the effort would stop. If the perceived benefits appear to outweigh the associated costs, the work effort would continue.

Second, it should develop a countertrade strategy. An important step involved in developing a countertrade strategy is to obtain top management support. Support for the countertrade effort is necessary from the top level of senior management as well as from key leaders and functional-level managers impacted by countertrade throughout the firm.

The major critical success factors to developing top management support for countertrade include:

- Creating the understanding that the fundamental mechanism of doing business internationally requires the use of various forms of countertrade.
- Identifying and describing the benefits, including competitive advantages, that would accrue to the multinational firm participating in a countertrade effort.
- Developing general plan of the broad-based requirements for the firm to move into the countertrading environment.
- Obtaining preliminary support from foreign companies or countries who would be willing to participate in the countertrade effort. This support would indicate to top management that there is a willing group of individuals who, on a joint basis, would like to foster a closer, nonmonetary trading partnership.

FIGURE 7-5
Contertrade Activity Process Flow

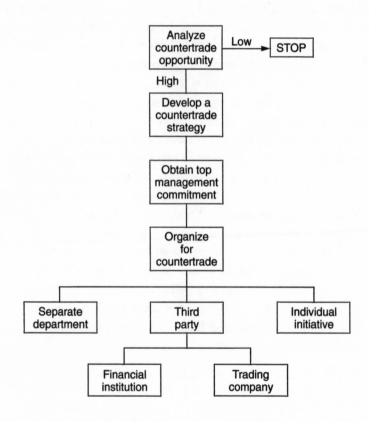

Third, the company should organize for the countertrade effort. Numerous alternatives exist regarding how a firm might perform the countertrade effort.1 Two principles that should be followed in organizing for the countertrade effort are (1) clear leadership of the effort must be established and (2) the effort should be multifunctional in scope. An emphasis on coordinating the appropriate functional inputs is very important.

At this phase in the countertrade process, the multinational corporation must decide whether it should handle the countertrade effort itself or select a third party to manage the effort for it. The company that decides to handle the countertrade effort in-house must decide whether to set up a separate countertrade department or handle each countertrade deal as an individual event. For the firm who expects to do more than just dabble in the countertrade arena, it is suggested that a separate countertrade department be set up. This department should consist of personnel from various functional areas such as marketing, purchasing, manufacturing, legal, and finance. An important task is to integrate the countertrade department within the company. There is no one best way to do this integration. Some firms have reorganized themselves in order to foster such an integration.

The company that decides to select a third-party provider to handle the countertrade effort must decide whether to use a trading exchange, a financial institution like a bank, or both. For the multinational company who is embarking upon countertrade for the first time, it is recommended that a third party be hired, especially one who has had significant experience.

Figure 7-6 provides a summary that compares the advantages and disadvantages of using a third-party trading company for countertrade activities versus performing the tasks in-house. The third party gives the multinational a great deal of countertrade experience, international contacts, an existing countertrade network, and legal and political acumen. Unfortunately, using a third party also provides the potential for impersonal service, self-interested advice, a lack of security and confidentiality, and a large commission.

On the other hand, performing the countertrade activity in-house provides for lower costs (no commission), greater control, increased flexibility in deal making, and direct contact with the countertrade partner. Performing countertrade in-house does have its disadvantages. It requires a close coordination of several functional departments, funding for the education and training of inexperienced personnel, and a drain on scarce managerial resources.

Negotiating Your Own Countertrade Deals

For the company that decides to negotiate its own countertrade deals, here are some helpful hints:

1. It is important that a multinational firm faced with a partner's demand to countertrade not overreact. Countertrade is legal and, at times, beneficial to both parties.
2. If possible, start the negotiations by offering an offset proposal, but do not initially make any firm commitment to trade goods.
3. Keep the ratio of the value of the goods taken in countertrade to the goods given out as low as possible.
4. Avoid deals where specific goods must be taken in countertrade. Try to negotiate a market basket of goods you could take in exchange.

FIGURE 7–6
Organizing for Countertrade

	Advantages	Disadvantages
In-house	Lower cost Control Flexibility Direct contact	Close department coordination Education and training Manpower needed
Third party	Countertrade expertise Contacts Network in place Legal acumen	Impersonal Self-interested Lack of confidentiality Commission

---

5. Elongate the time frame over which the countertrade obligation must be fulfilled. The average arrangement is for one to three years, but there have been some arrangements for one or two decades.
6. Specify quality, delivery, and warranty requirements explicitly. Determine each party's recourse for non-performance.
7. Determine any restrictions on resale territory for countertrade goods. Try to keep these limitations as minimal as possible for your firm.
8. Structure the countertrade deal so that you can get as much cash as possible instead of goods.
9. Try not to take unfamiliar products or commodities as part of the countertrade deal. Stick with products that you understand.
10. Avoid firm export prices unless you understand the intricacies of hedging for exchange rate fluctuations.
11. Find out who actually has the authority to authorize a deal. This can be a big problem in many centrally planned economies.
12. Remember, the Department of Commerce states that only one out of four countertrade negotiations are ever completed.

Government's Position on Countertrade

The U.S. government does not involve itself in countertrade agreements unless such agreements violate U.S. trade laws or involve national security, or government involvement is requested by a U.S. multinational corporation. This nonintervention policy does not mean that the government condones countertrade obligations. On the contrary, the U.S. government views countertrade as a restrictive and inefficient trade policy, counterproductive to a market-oriented trading system free of restrictions. The U.S. government as a matter of public policy vehemently opposes foreign countertrade arrangements that were mandated by statute or regulation.

The U.S. Government feels that countertrade arrangements discriminate against small firms since the countertrade process is quite complicated and expensive. It is also difficult for a small firm to turn countertraded goods for which they have no internal use into hard currency. Many small firms lack the resources to dedicate several individuals to countertrade arrangements on a full-time basis.

If requested by a U.S. multinational corporation, the government will formally protest any countertrade arrangement mandated by one of our foreign trading partners. The government presently has several countertrade protests and consultations in effect with many of our trading partners. Unfortunately, these protestations will not stem the increase in countertrade agreements. In fact, the U.S. government itself participates in several countertrade arrangements. The Agriculture Department and the General Services Administration both have negotiated countertrade arrangements with foreign trading partners. Today, the government has taken, by necessity, a more pragmatic view of the international marketplace and does not automatically oppose all countertrade involvement by either itself or U.S. multinational corporations. This does not mean that the Internal Revenue Service overlooks countertrade involvement by U.S. companies.

The tax laws relating to international countertrade are not totally clear. The IRS would like to apply the same treatment to a countertrade arrangement that it does to a cash transaction, but this is not always possible. Countertrade arrangements have become so intricate and lengthy that setting a dollar value to these arrangements is complex and difficult. The IRS states that the fair market value of the goods and services countertraded is considered gross income. But this fair market value is often difficult to estimate since countertraded goods are frequently those that cannot be sold in the open market and the value of these countertraded goods is often inflated for the purposes of the countertrade negotiation.

Nevertheless, in 1982 Congress passed the Tax Equity and Fiscal Responsibility Act (TEFRA). This law placed strict reporting guidelines on third party trade exchanges. These exchanges must report all transactions that result in client income to the Internal Revenue Service (IRS). The exchanges also report the equivalent dollar value of trade credits to the IRS as income. Therefore, it is mandatory that all U.S. multinational corporations report their countertrade income to the IRS. The problem is setting a fair market value to countertrade arrangements for reporting purposes.

ROLE OF PURCHASING

Closely following the advent of countertrade, strong corporate interest in offshore sourcing has emerged. Many companies have assigned this task to the international materials management function, which incorporates foreign purchases as part of an overall strategic plan.

Countertrade arrangements triggered by transactions that conform to the international economy's trading rules can be integrated into the multinational corporation's overall strategic plan for global sourcing only if the worldwide purchase of goods and services is included in the plan. Ideal sourcing opportunities through countertrade can be found in countries that are just beginning to impose countertrade requirements and where competition for foreign sources of supply are just beginning. A problem that gets worse in a strong economy is that suppliers of popular products become increasingly reluctant to countertrade. In such situations, a company may have to settle for dealing with their second or third choice for a supplier rather than their first choice.

Perhaps the primary reason multinational companies shift a share of their purchases to off-shore sources is to satisfy countertrade obligations in customer countries where sales transactions have been completed or may be placed in the future. Many multinationals view countertrade as the catalyst to generate future sales opportunities. Forging strong supplier linkages before exploiting sales opportunities within a foreign country requiring countertrade obligations can encourage solutions and avoid problems resulting from potentially unreasonable demands by customers.

On the other side of the countertrade partnership, countries which link countertrade requirements with imports understand (1) the benefits of developing long-term relationships with multinational corporations; (2) the importance and availability of technology transfer from industrialized nations; (3) the need for improved manufacturing techniques in order to become globally competitive; and (4) the resulting increased employment and economic development. These lesser-developed countries see countertrade as the best method available to obtain a permanent position in the multinational corporation's global materials network.

It is clear that the worldwide system of countertrade requires a global purchasing strategy. It also encourages longer-term partnerships and the transfer of technology, quality controls, and manufacturing skills from industrial countries to lesser-developed nations. In the final analysis, it fosters an environment where a premium is placed upon efficient and timely purchasing.

Strategic Sourcing

With multinational corporations fast becoming global trading companies as well as manufacturing firms, the purchasing function, as an integral part of a firm's global materials sourcing strategy, is being asked to expand its traditional role in material procurement and perform a wide variety of tasks, not all of which are related to its traditional function. For example, 11 percent of the Fortune 500 companies responding to a *CPI Purchasing* magazine survey reported that the purchasing function was involved in selling goods accepted in countertrade deals for which the company could find no internal use. More importantly, 27 percent of these same companies stated that the purchasing department initiated contacts with prospective trading partners. In these cases, purchasing first identified potential sources of supply in countries with mandatory countertrade laws and then involved the marketing function, a clear reversal of the conventional countertrade roles. Historically, marketing would locate a potential market for the goods manufactured by the multinational and consummate a countertrade arrangement in order to complete the sale. Purchasing would subsequently become responsible for disposing of the goods taken in countertrade through either internal use or external sale.

For example, one multinational company developed a policy called "reverse countertrade." Company representatives began the countertrade arrangement by presenting their foreign trading partners with their present and future sourcing requirements. Only after the representatives had established the availability of these goods and the feasibility of purchasing them did they get around to the multinational's main objective of selling their own goods abroad.

The purchasing function is also involved in many traditional activities applied to countertrade arrangements. For instance, many multinational corporations make use of purchasing's expertise in negotiations when pounding out countertrade arrangements with foreign trading partners. In the survey performed by the authors, 54 percent of the respondents reported that purchasing negotiated countertrade contracts; 49 percent reported that purchasing set the monetary value of goods accepted in countertrade; 46 percent reported that purchasing found internal uses for countertrade materials; 43 percent reported that purchasing determined the quality of goods accepted in countertrade; and 41 percent reported that purchasing conducted on-site visits of foreign production facilities of potential countertrade partners. Figure 7–7 graphically summarizes the roles played by the purchasing function in international countertrade.

Purchasing's involvement with countertrade arrangements has for the most part, impacted favorably on international sourcing. When involved early in the process, purchasing has discovered that countertrade can lead to low-cost sources of supply. It is not uncommon for a countertrade arrangement to result in a beneficial relationship that lasts long after the countertrade arrangement is completed. In the *CPI Purchasing* survey, 53 percent of the companies surveyed said that they continued to source from suppliers after the countertrade arrangement was completed. Not surprisingly, these same firms were, for the most part, the ones in which purchasing played a prominent countertrade role. Purchasing personnel, by training, tend to look beyond the one-time buy and view sourcing relationships as being potentially long-term. As mentioned earlier, foreign trading partners are, for the most part, quite willing to foster long-term relationships with multinational corporations by giving favorable countertrade terms and conditions.

A major role played by purchasing is to ensure the quality of goods and services purchased from vendors, regardless of whether these vendors are foreign or domestic. In many instances, purchasing works closely with critical domestic suppliers in order to maintain or improve the quality of incoming materials. This relationship for quality improvement does not seem to exist to the same extent for foreign countertrade partners as it does for domestic sources of materials. In the *CPI Purchas-*

ing survey, 30 percent of the respondents said that purchasing worked with countertrade partners while setting initial product specifications, but only 5 percent of the respondents said that purchasing worked with these countertrade partners to maintain or improve the quality of these goods after the arrangement went into effect.

Involve Purchasing Early

Even though it is not a prerequisite for the purchasing function to be involved early in the countertrade process, some multinational corporations' countertrade arrangements have failed because purchasing was not involved until it was too late. For example, one multinational corporation concluded, without the assistance of purchasing, a $90 million countertrade arrangement with a foreign trading partner. Subsequently, the multinational asked its purchasing function to dispose of

---

FIGURE 7–7
International Countertrade

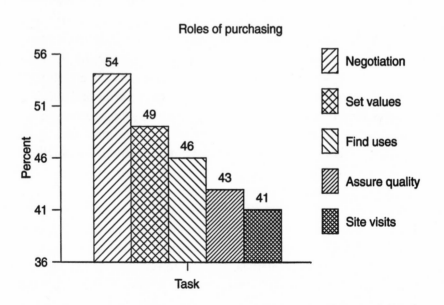

the goods taken in this countertrade arrangement. The purchasing function informed management that the goods taken in countertrade were of inferior quality, could not be used internally, and must be sold at a loss.

This example highlights the strong qualities that purchasing can bring to countertrade arrangements. While the marketing function may agree to almost anything to consummate a sale, purchasing professionals are far more inclined to keep the traditional criteria of price, quality, delivery, and service in mind while evaluating and negotiating countertrade arrangements. Many insightful companies are adopting a policy forbidding countertrade arrangements without the involvement and final approval of the purchasing department.

Should purchasing take a more active role in countertrade arrangements, including both negotiations with and development of countertrade partners? The answer is yes for several reasons. First, cost reduction has become a major corporate goal, and countertrade can be a major source of low-cost materials and services. Second, multinationals that find and develop foreign sources of quality materials will have a strategic advantage over their slower competitors who may be stuck with what is left. Third, for the multinational corporation, countertrade is here to stay.

LESSONS FOR MANUFACTURING

An eclectic comparison of Figures 7–3 and 7–4 demonstrates a trend in countertrade arrangements that has serious implications for the multinational corporation. In the *CPI Purchasing* 1986 Fortune 500 survey, offsets were the most common type of countertrade arrangement, practiced by 73 percent of the multinational companies. This is a significant increase from the 1983 survey. This increase in the practice of offset trading along with the decrease in the percentage of multinational firms involved in switch trading may be an indication that multinational corporations are getting more directly involved in countertrade arrangements and that this involvement is more long-term in nature. More fundamentally, the trend in countertrade seems to be away from one-time, short-term arrangements toward longer-term comaker relationships.

Multinationals are looking for more permanent foreign sources that can be developed into competitive extensions of their global sourcing strategy.

This trend in countertrade is not unexpected. As mentioned previously, many countries view countertrade as an internal economic development strategy. With the advent of the debt crisis for third world countries, money from public sources is very difficult to obtain. Private investment by multinational corporations through countertrade mechanisms has become the most viable means for third world countries to obtain investment capital.

Multinational corporations also view a longer-term relationship with third world countries as beneficial. First, these countries can be a prime source of low-cost labor and materials. Second, these same countries have relatively untapped markets for many goods marketed by multinational firms.

The switch from short-term arrangements, where manufactured goods are traded for commodities and cash, to longer-term supplier investment strategies has significant implications for the approach taken towards countertrade in the future. Multinational corporations must now view their countertrade partners as an important part of their overall corporate strategies rather than as just a source of questionably useful goods.

What then should be the strategic objective of countertrade? Should a multinational corporation view countertrade as merely a necessary evil in order to market their products abroad? Insightful multinational companies are beginning to view countertrade as more than just a traditional and potentially disagreeable function. These multinationals are using countertrade as another competitive weapon in their strategic sourcing arsenal. They are moving beyond the reactionary practices of finding uses for goods no one really wanted and probably could not use. The innovative multinational companies are taking an active approach to countertrade arrangements. These multinationals have reversed the traditional countertrade process and are first looking for market opportunities and then identifying potential countertrade partners who can best help them exploit these opportunities.

ENDNOTES

1. At this juncture, it is also necessary to decide whether the multinational firm wants to countertrade domestically, internationally or both. Unknown to many multinational firms, non-monetary trade is very popular domestically. The opportunity for countertrade may, for any specific company, be just as great in the United States as it is elsewhere.

CHAPTER 8

MANAGING FOREIGN EXCHANGE RATES (ADVANCED TOPICS IN PURCHASING)

INTRODUCTION[1]

The importance of the purchasing function to the overall competitiveness of the firm has become more evident in recent years. For many firms, a sizable percentage of the total cost of their products is accounted for by purchased components, and this percentage is increasing. General Motors Corporation has estimated that it spends in excess of 58 percent of its sales dollars on purchased components, not including transportation costs.

These large expenditures place increasing pressure on purchasing operations to obtain components at the lowest possible cost without sacrificing quality and service. To accomplish this objective, purchasing managers are eager to locate low-cost, technically capable, high-quality sources of supply wherever they may exist. U.S. corporations are increasingly turning to nondomestic suppliers to satisfy these constraints. Purchasing managers have discovered that the cost of many foreign goods (including transportation) is simply lower than their domestically produced counterparts of comparable quality, even allowing for premiums associated with possible warehousing requirements.[2]

The medium of payment for most foreign purchases is currency.[3] The currency used for foreign purchases may be the buyer's, the supplier's, or possibly the currency of a third country. The currency selected for payment can result in higher or lower purchase costs over the life of a contract due to volatile exchange rates. This chapter explores foreign exchange rate issues in international purchasing and presents a framework of exchange rate management techniques available to U.S. manufacturers purchasing abroad.

THE IMPORTANCE OF EXCHANGE RATES IN INTERNATIONAL PURCHASING

Exchange rates impact the price paid for imported merchandise when payment is in the supplier's currency and there is a lag between when the contract is signed and when payment is made. Depending upon the country of the supplier and the direction of the exchange-rate movement, a buyer can end up paying substantially more or less than the original contract price. For example, suppose that on January 3, 1986, a U.S. buyer enters into a one-year contractual agreement with a Japanese supplier that calls for equal monthly payments in yen. Furthermore, assume that the contract, when signed on January 3, specified that at the prevailing exchange-rate, 202.48 yen per dollar, each monthly payment would equal 20.248 million yen ($100,000 U.S.). Depending upon the exchange rate volatility of the yen with respect to the U.S. dollar over the duration of the contract, the total cost to the buyer could be higher or lower than the expected $100,000 per month.

Table 8-1 presents data supporting an analysis of expected U.S. dollar expenditures over the 12-month life of this contract. Since the U.S. dollar lost buying power against the yen during 1986, the contract with the Japanese supplier would have cost a total of $1,433,230 U.S. over the twelve-month period—$233,230 U.S. more than the $1.2 million originally expected.

The problem becomes even more complicated when two or more foreign suppliers from countries using different currencies are under consideration for a contract. Each potential supplier's currency would fluctuate differently in relation to the U.S. dollar. Therefore, each contract, if consummated, would ultimately cost different U.S. dollar amounts. These examples illustrate the impact that exchange rates can have on the cost of imported goods and demonstrate the importance of considering exchange rates in sourcing decision making.

TABLE 8–1
U.S. Dollar Contract Expenditures

Month	Projected Yen Expenditures	Exchange Rate*	Actual U.S. Dollars Needed
1	20.248 million	202.48	$ 100,000.00
2	20.248 million	192.30	105,293.80
3	20.248 million	180.15	112,395.20
4	20.248 million	180.45	112,220.80
5	20.248 million	169.63	119,365.60
6	20.248 million	174.10	116,300.90
7	20.248 million	161.10	125,685.90
8	20.248 million	153.65	131,780.00
9	20.248 million	155.50	130,212.20
10	20.248 million	154.30	131,224.80
11	20.248 million	163.50	123,840.90
12	20.248 million	162.10	124,910.50
		Total	$1,433,230.60

* Yen/U.S. dollar.

RESPONDING TO CURRENCY MOVEMENTS

There are various approaches a firm can pursue in response to volatile exchange rates. Domestic buyers making payment in a foreign currency may attempt to lessen the risk of an adverse price fluctuation through the use of currency futures or a risk-sharing contractual agreement. On the other hand, a firm may wish to exploit existing or expected favorable exchange rate differentials in identifying and selecting suppliers. For example, a firm might deliberately place orders with suppliers whose currencies are expected to lose value against the dollar.

There are factors involved with currency fluctuations that can be recognized and tracked over time. Exchange rates move in cycles. In the past, attempts to forecast short-term exchange rates have not met with much success; however, some technically-oriented forecasting services have done well in forecasting longer-term directional currency movements.[4] Longer-term movements are becoming more predictable as the value of a country's currency becomes a means of fostering increased exports, decreased imports, or both. For example, the drop in the value of the dollar in relation to other currencies (e.g., the Japanese yen and West German mark) during 1986 was the result of an explicit policy implemented by the United States and other governments. The knowledgeable purchasing manager either did not generate requirements contracts payable in these currencies or found a way to moderate the impact of the declining value of the dollar.

Exchange-Rate Management: A Conceptual Framework

Research has identified a wide range of currency management approaches. This has led to the construction of a conceptual framework for exchange-rate management approaches. The conceptual framework for currency management strategies is principally designed for U.S. importers serving domestic markets. Companies with both domestic and international markets have additional strategies and resources available to them, some of which are beyond the scope of this chapter.

The conceptual framework consists of two major strategy categories: the *macro level* and the *micro level*. Macro-level strategies affect the sourcing decision and the volume/timing of purchases; micro-level strategies are employed after both the source selection and volume-timing decisions have been made and are frequently used to protect the buyer from adverse currency fluctuations.

The conceptual framework for both macro- and micro-level exchange rate management is presented in Table 8-2.

USING EXCHANGE RATES IN THE SUPPLIER SELECTION DECISION

The use of exchange-rate information in the supplier selection decision is important to successful global sourcing. Because of volatile exchange rate fluctuations, two equally capable suppliers from different countries who quote the same U.S. dollar equivalent price at a particular point in time can end up having significantly different prices over the extended life of a requirements contract.

TABLE 8–2
Conceptual Framework for Currency Management

Macro Strategies

1. Exchange rate information is used as an input to the source selection decision.
2. Exchange rate information is used as an input to the volume-timing of purchases decision.

Micro Strategies

1. Payment in U.S. dollars.
2. Buying foreign currency forward.
3. Buying foreign currency futures.
4. Risk-sharing contract agreement.
5. Payment in supplier currency.

Purchasing managers are trained to thoroughly screen potential foreign suppliers using a myriad of complementing factors which include quality, delivery, service, technical support, price, financial status, and managerial capability. But it is still rare to find books, articles, and trade publications that simultaneously explain the importance of exchange rates in international sourcing and advocate their use in the supplier selection decision.

Supplier Selection: An Example

At the beginning of August 1986, company X has a choice of entering into a requirements contract for an electronic subassembly with either of two equally qualified suppliers, one Japanese and one Korean. The contract with the Japanese supplier calls for the payment of yen in 22 biweekly installments of 15.365 million yen ($100,000 U.S.); the contract with the Korean supplier calls for the payment of won in an equal number of biweekly installments of 88.41 million won ($100,000 U.S.).

On August 1, 1986, the value of these two currencies in relation to the dollar was 153.65 yen per dollar and 884.1 won per dollar. If the values of these two currencies with respect to the dollar remain constant over the payment horizon, company X would pay $2 million U.S. (22 biweekly payments of $100,000 U.S. each) for its purchases, regardless of which supplier it selected for the order. However, given that the values of the yen and won with respect to the dollar will most likely fluctuate over time, company X is uncertain which supplier to select.

Figures 8-1 and 8-2 show the actual exchange rates for the Japanese yen and Korean won, respectively, from August 1, 1986 through May 25, 1987. This time period covers the 22 biweekly periods of the requirements contract mentioned above. These figures demonstrate that the value of these currencies did fluctuate over the contract period; thus, the actual cost of the requirements contract with either supplier did not equal $2 million U.S. Not only was the actual cost of the contract to company X different than the budgeted amount of $2 million U.S., but this actual amount was not equal for the two suppliers because of differences in the way the value of the two currencies fluctuated in relation to the U.S. dollar over time.

Table 8-3 contains the actual expenditures in dollars for company X if it had placed monthly order releases to the Japanese supplier. Under this scenario, company X would spend $2,209,293.70 U.S. for the 22 biweekly payments which is slightly more than the budgeted $2,200,000.00 U.S. In contrast, Table 8-4 contains the same actual expenditures, expressed in dollars, for the Korean supplier. Contracting with the Korean supplier, company X would spend $2,256,263.00 for the same 22 biweekly payments. Not only is this amount significantly higher than the budgeted U.S. dollar amount of $2,200,000.00 U.S., but it is $46,969.30 U.S. ($2,256,263.00 − 2,209,293.70) higher than would have been paid to the Japanese supplier. Clearly, expected exchange-rate movements should have played a significant role in deciding which of the two suppliers to select.

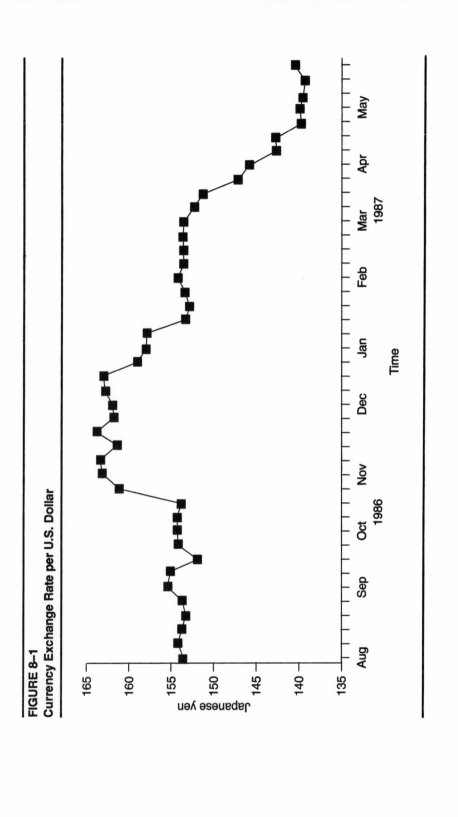

FIGURE 8–1
Currency Exchange Rate per U.S. Dollar

FIGURE 8–2
Currency Exchange Rate per U.S. Dollar

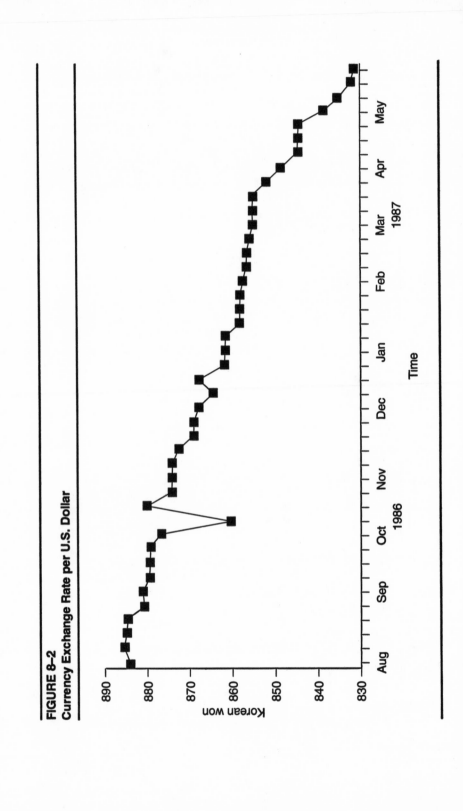

TABLE 8–3
Japanese Yen

Order Date	Yen Needed for Purchases (millions)	Exchange Rate (Yen/Dollars)	U.S. Dollars Needed
1. August 1, 1986	15.365	153.65	$100,000.00
2. August 15	15.365	153.87	99,857.02
3. August 29	15.365	153.80	99,902.47
4. September 12	15.365	155.25	98,969.40
5. September 26	15.365	154.30	99,578.74
6. October 1	15.365	154.40	99,514.25
7. October 24	15.365	161.40	95,198.27
8. November 7	15.365	163.50	93,975.54
9. November 14	15.365	164.00	93,689.02
10. December 5	15.365	162.10	94,787.17
11. December 19	15.365	163.10	94,206.01
12. January 2, 1987	15.365	158.10	97,185.33
13. January 16	15.365	153.40	100,162.97
14. January 30	15.365	153.50	100,097.72
15. February 13	15.365	153.60	100,032.55
16. February 27	15.365	153.66	99,993.49
17. March 13	15.365	152.30	100,886.41
18. March 27	15.365	147.20	104,381.79
19. April 10	15.365	142.80	107,598.04
20. April 24	15.365	139.75	109,946.33
21. May 8	15.365	139.55	110,103.91
22. May 22	15.365	140.67	109,227.27
	Total		$2,209,293.70

Situation 1: Supplier Selection at Eaton Corporation

Eaton Corporation employs quantitative models developed by its corporate economists to forecast yearly average inflation rates and yearly average exchange rates for various countries and currencies of interest. These forecasts are used for predicting the yearly average percentage changes in a foreign country's prices relative to U.S. prices. This information is one of many inputs the company uses in making sourcing decisions.

For example, in late 1986, the following comparison of West German price changes relative to the United States for 1987 was issued to

TABLE 8–4
Korean Won

Order Date	Won Needed for Purchases (millions)	Exchange Rate (Won/Dollar)	U.S. Dollars Needed
1. August 1, 1986	88.41	884.1	$100,000.00
2. August 15	88.41	884.8	99,920.89
3. August 29	88.41	880.7	100,386.06
4. September 12	88.41	879.3	100,545.89
5. September 26	88.41	879.1	100,568.76
6. October 1	88.41	860.0	102,802.33
7. October 24	88.41	874.0	101,155.61
8. November 7	88.41	874.0	101,155.61
9. November 14	88.41	868.9	101,749.43
10. December 5	88.41	867.5	101,913.54
11. December 19	88.41	867.3	101,937.05
12. January 2, 1987	88.41	861.4	102,635.24
13. January 16	88.41	858.1	103,029.95
14. January 30	88.41	857.9	103,053.97
15. February 13	88.41	856.3	103,246.53
16. February 27	88.41	855.8	103,306.85
17. March 13	88.41	855.0	103,403.51
18. March 27	88.41	851.8	103,791.97
19. April 10	88.41	844.4	104,713.96
20. April 24	88.41	844.3	104,713.96
21. May 8	88.41	835.2	105,854.89
22. May 22	88.41	831.1	106,377.09
	Total		$2,256,263.00

Eaton purchasing managers (see Table 8–5). The report indicated that West German prices would jump 17.3 percent compared to U.S. prices in 1987.

Information of this type has proven valuable to Eaton Corporation in making sourcing decisions. A prime example is the case of iron castings. In 1984 and 1985, Eaton was sourcing iron castings in Europe and the Far East. The original impetus for sourcing offshore was the strength of the U.S. dollar at this time. However, in late 1985 and early 1986, Eaton's exchange rate forecasts indicated a downward trend in the buying power of the U.S. dollar against the European and Far Eastern currencies of interest over a five-year forecast horizon. The result was

TABLE 8-5
**1987 Comparison of West German Price Changes Relative to
the United States**

(A) Forecast 1987 Inflation West Germany	(B) Forecast 1987 Inflation U.S.	(C) Ratio [A+100]/[B+100]	(D) 1986 Exchange Rate (DM/U.S.$)	(E) Forecast 1987 Ex. Rate (DM/U.S.$)	(F) Ratio (E/D)	(G) Average Percent Change (C/F-1) × 100
1.537	2.313	0.992	2.163	1.830	0.846	+17.309

an upward trend in forecast price changes for these countries compared to U.S. prices (see column (G) in Table 8-5, for example) for the same time horizon. Forecasts for other countries of interest also indicated a weakening in the buying power of the U.S. dollar over a five-year planning horizon.

Since exchange-rate forecasts indicated a weakening of the U.S. dollar against a wide range of foreign currencies, it did not make sense for Eaton to move its iron castings to different foreign sources. Instead, Eaton brought the iron castings back to its domestic sources and was able to maintain a worldwide competitive price. One reason for this was the stiff global competition challenging the domestic sources in the interim period, causing them to realize their need to become more cost-efficient. To become competitive, they invested extensively in their production processes and, as a result, were attractive source candidates at the time Eaton Corporation was reexamining available sourcing alternatives.

USING EXCHANGE RATES IN THE VOLUME-TIMING OF PURCHASES

Historically, purchasing has viewed markets as being dichotomous— those markets in which the factors of supply and price could be considered reasonably stable in the short run and those markets in which the factors of supply and price fluctuate significantly in the short run. For example, products purchased in stable markets might consist of standard off-the-shelf items such as many maintenance, repair, and operating supplies purchased from domestic sources. Because the price of such materials is relatively stable over the short term, the volume-timing of purchases in order to exploit favorable price differentials is a nonissue.

In contrast, unstable markets provide the buyer with an opportunity to either exploit favorable price changes or avoid unfavorable price differentials through the timing of purchases. If the price of a particular item is expected to increase, the buyer could purchase in a larger-than-usual quantity and store the item in inventory until needed. As stated by Dobler, Lee, and Burt:

> Timing [of purchases] is a much more important matter when a purchase is made in a market which tends to be unstable. Careful observation and analysis of market conditions are essential if buyers hope to satisfy their objectives relative to supply and price. Although they cannot influence the *market price,* they can, by their timing of purchases, control to some extent the *price they pay.* In a highly competitive business, the results of their timing can significantly affect both the competitiveness and profitability of their firm's operation [Emphasis supplied.].[5]

In the international marketplace, there really is no such thing as a stable marketplace. Even though a foreign supplier consistently quotes the same price for his products over time, the exchange rate for that supplier's currency with respect to the dollar is constantly changing. In such cases, the timing of purchases can significantly impact the material's purchase cost, final product price, and thereby, overall product competitiveness.

Timing Purchases: An Example

Returning to the supplier selection example presented earlier, let's assume that after careful analysis, company X decides to place the requirements contract with the Japanese supplier instead of the Korean supplier because based on forecast information, it anticipates a smaller loss in the buying power of the U.S. dollar against the yen than against the won in future periods. The anticipated schedule of payments to the Japanese supplier was given earlier in Table 8-3.

As pointed out earlier, the value of the yen in relation to the dollar was 153.65 yen per dollar on August 1, 1986. If the value of the yen remained constant for the 44-week period (22 biweekly payments), company X would pay $2 million U.S. for its purchases. This is company X's budgeted purchase cost for its requirements contract. In reality, as demonstrated in Figure 8-1 the value of the yen fluctuated substantively over the 44-week period, with the yen becoming consistently more expensive after December of 1986.

Referring to Table 8-3 once again, one can see that the actual expenditures in dollars for company X, if it placed 22 biweekly order releases to the Japanese supplier, was $2,209,293.70. Could company X have done better if it used the volume-timing of purchases technique?

The volume-timing of purchases technique uses a forward buying policy when the value of the yen is increasing in relation to the dollar and a short-term purchasing policy when the value of the yen is decreasing in relation to the dollar. Suppose in December, company X experienced a pronounced decline in the buying power of the U.S. dollar with respect to the yen. Furthermore, assume that available forecast information indicated the decline would continue. At this juncture, company X could decide to implement a forward buying policy to protect itself against the falling value of the dollar. One such policy alternative is pictured in Figure 8-3 which shows the same currency exchange curve referenced previously, but with arbitrary purchase periods imposed upon it. Table 8-6 presents the expenditures in dollars for company X if it follows this particular forward buying plan.

Under the plan presented in Table 8-6, company X places biweekly orders with the Japanese supplier from August 1 through January 1, 1987, as the dollar reacts favorably in relation to the yen. From January 2, 1987 through May 31, 1987, as the yen consistently increases in value in relation to the dollar, company X places two order releases with the Japanese supplier, one on January 2, 1987, and another on March 20, 1987. The empirical results from the volume-timing of purchases is a savings of $34,595.30 ($2,209,293.70 - 2,174,698.40) over the standard biweekly purchase plan. If we allow for the increase in inventory holding costs[6] caused by the volume-timing of purchases, the net savings equals $10,707.80.

Whether other purchase plans made using the volume-timing technique yield greater or lesser savings depends upon the timing of the

FIGURE 8-3
Currency Exchange Rate per U.S. Dollar

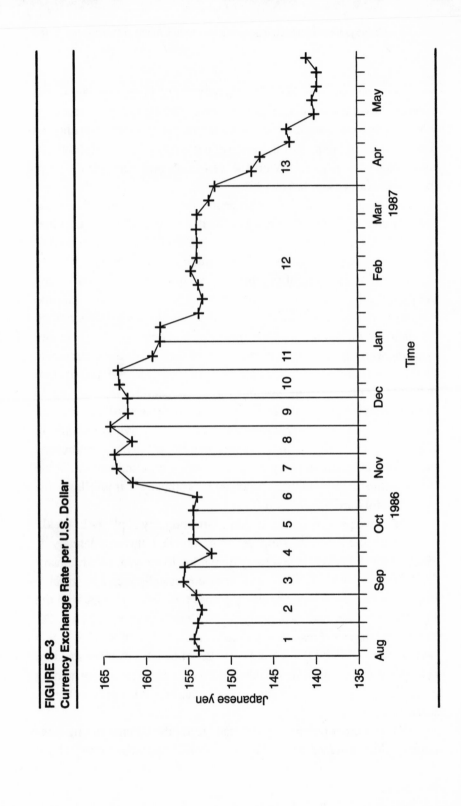

TABLE 8–6
Japanese Yen

Order Date	Yen Needed for Purchases (millions)	Exchange Rate (Yen/Dollar)	U.S. Dollars Needed
		1986	
1. August 1	15.365	153.65	$ 100,000.00
2. August 15	15.365	153.87	99,857.02
3. August 29	15.365	153.80	99,902.47
4. September 12	15.365	155.25	98,969.40
5. September 26	15.365	154.30	99,578.74
6. October 10	15.365	154.40	99,514.25
7. October 24	15.365	161.40	95,198.27
8. November 7	15.365	163.50	93,975.54
9. November 14	15.365	164.00	93,689.02
10. December 5	15.365	162.10	94,787.17
11. December 19	15.365	163.10	94,206.01
		1987	
12. January 2	92.190 million*	158.10	$ 583,111.95
13. March 20	76.825 million†	147.20	521,908.96
		Total	$2,174,698.40

*A twelve-week purchase period.
†A ten-week purchase period.

purchases, the currency in question, and the volatility of the exchange-rate movements. The amount of actual benefit is also dependent upon the buyer's ability to predict directional trends in the movement of the exchange rate. The more accurate the forecast, the greater the savings potential. For this reason, the importance of accurate forecasting should not be underemphasized.

Situation 2: Volume-Timing of Purchases at Conoco Inc.

Conoco Inc. is one of the world's largest petroleum companies. The company sources a significant amount of its nonpetroleum material needs internationally. These foreign sourcings are predominantly pipes, valves, and fittings designed for use in oil exploration, production, and processing.

In April 1986, Conoco was faced with a sourcing situation for seamless tubular pipe where the prices paid to Japanese mills, a key foreign source, was forecast to increase substantially in the near future. This forecast price increase was the result of two major changes in the economic environment. First, the Japanese seamless tubular steel industry had been experiencing significant losses in sales resulting from depressed demand worldwide.[7] Second, the value of the U.S. dollar in relation to the Japanese yen had recently been decreasing, and Conoco forecast a continued downward trend for the value of the dollar in relation to the yen.

After a careful analysis of the situation, Conoco decided to commit early for a one-year requirement of seamless tubular pipe at fixed prices recognizing the severe foreign upward price pressure caused by the continued devaluation of the U.S. dollar in relation to the Japanese yen. This volume-timing of purchases for seamless tubular pipe resulted in a significant ($1 million U.S.) monetary savings for Conoco.

MICRO–LEVEL EXCHANGE RATE MANAGEMENT STRATEGIES

Micro-level strategies can be employed after both the source selection and volume-timing of purchases decisions have been made. These strategies are frequently used to protect the buying firm from unexpected, shorter-term currency fluctuations.

Payment in U.S. Dollars (Risk Avoidance)

One way to avoid the risk of volatile exchange rates is for the buyer to pay in U.S. dollars. This strategy transfers the risk of an adverse currency fluctuation from the buyer to the supplier. Consider the following example. Assume that a contract calls for the payment of 983,808 U.S. dollars instead of 2,418,200 German marks in six months.[8] By specifying payment in U.S. currency, the buying firm insures itself a fixed future

TABLE 8–7
Cost in U.S. Dollars for Three Planning Cases

	Buying Power of the U.S. Dollar		
Strategy	Decreases by 10 Percent (2.2122 DM per $1)	Remains Unchanged (2.4580 DM/$1)	Increases by 10 Percent (2.7038 DM per $1)
Pay in U.S. dollars	$ 983,808	$ 983,808	$ 983,808
Baseline approach	$1,093,120	$ 983,808	$ 894,371

price in U.S. dollars regardless of whether the exchange rate fluctuates. Thus, the buyer has a firm cost for planning purposes and has precluded the occurrence of future surprises.

Table 8-7 compares this strategy with the baseline approach of paying in the supplier's currency with no protection. In the event of a 10 percent decrease in the buying power of the U.S. dollar, the strategy saves the buying firm from a loss of $109,312 U.S. On the other hand, if the buying power of the U.S. dollar increases by 10 percent, the buyer misses a potential gain of $89,437.

Buying Foreign Currency "Forward" (Risk Avoidance)

An alternative strategy is to buy "forward," for future receipt, the needed foreign currency from a financial institution at a fixed exchange rate. This practice also removes the risk of an unfavorable currency fluctuation, providing the buyer with a firm future price. The forward market for foreign exchange is a worldwide network consisting mainly of banks and brokers trading electronically. The amount of currency covered by a given contract is determined by negotiation and quotations are for a stated number of days into the future, based on an individual's needs (e.g., 30, 90, or 180 days). In forward buying, the bank handling the transaction is paid a commission, which is set by the spread between the bank's buy and sell price; this commission is not easily determined by the customer. To engage in forward buying, the firm must have a line of credit with the bank. Because the

TABLE 8–8
Cost in U.S. Dollars For Three Planning Scenarios

	Buying Power of the U.S Dollar		
Strategy	Decreases 10 Percent (2.2122 DM/$1)	Remains Unchanged (2.4580 DM/$1)	Increases 10 Percent (2.7038 DM/$1)
Buying Forward	$1,000,207	$1,000,207	$1,000,207
Baseline Approach	1,093,120	983,808	894,371

expense of individual contracting is large, the forward market is limited to very large customers dealing in foreign trade.

Returning to the previous example, assume that payment is designated in the supplier's currency. Under this approach the buying firm purchases forward (180-day) 2,418,200 German marks. The exchange rate on January 3 is 2.4580 DM per U.S. dollar, and the published 180-day forward rate (as published in the *Wall Street Journal*) is 2.4182 DM per U.S. dollar. The forward rate quotation from the buyer's bank is estimated to be five points (0.0005 DM per U.S. dollar) lower than the published forward rate (i.e., 2.4177 DM per U.S. dollar). Thus, 2,418,200 marks bought 180 days forward costs $1,000,207 U.S. Since payment for currency bought forward does not become due until the firm receives the currency, the transaction has no opportunity cost.

Table 8-8 provides a comparison of this risk avoidance strategy with the baseline case under three planning scenarios. Note that if the buying power of the U.S. dollar decreases by 10 percent, the strategy results in a savings of $92,913. In contrast, if the buying power of the U.S. dollar rises by 10 percent, the strategy results in a loss of $105,836.

Purchasing Futures Contracts (Risk Minimization)

Another hedging mechanism is to buy a "future" or contract to purchase the needed amount of foreign currency from one of the currency commodity exchanges. If the buyer's selected currency of payment goes up in value in the interim before the payment becomes due, the buyer sells the contract for a profit and uses the profit to make up the difference when he makes payment. If, on the other hand, the buyer's selected currency falls in value, the buyer loses value when he sells the futures contract, but makes up most of the differential at the time of payment. Since realized returns on the futures contract are uncertain, this hedging mechanism does not remove risk altogether but it does reduce it. A detailed description of the remaining risk is contained in Appendix A.

Futures contracts are traded on an organized exchange, such as the Chicago Mercantile Exchange. While only individuals may be members of the exchange, brokers are afforded the opportunity to trade for their clients or for their own accounts. Unlike the forward market, which is limited to very large customers, the futures market is accessible to anyone needing hedge facilities. In the futures market, contracts are traded only on major currencies (German mark, British pound, Canadian dollar, Swiss franc, Japanese yen, and the Mexican peso), and the contract amounts and delivery or maturity dates are standardized. Maturity dates for currency futures occur on the third Wednesdays of March, June, September, and December. Fewer than 1 percent of the contracts are settled by actual delivery.

Suppose a firm decides to hedge its risk by buying a currency future. Recall the earlier example in which 2,418,200 German marks are required 180 days from January 3. Using this hedging method the firm purchases a contract to buy a set number of marks. The standard contract size for the German mark is 125,000 marks. Thus, the firm purchases 20 contracts of this size (2,418,200 divided by 125,000 rounds up to 20) on Friday, January 3, 1986. The settlement price of the day for a September[9] futures contract is 0.4150 dollars per mark. The cost of a single contract using this price is $51,875 (125,000 marks × $0.4150) and for twenty contracts, $1,037,500 ($51,875 × 20). This is the value of the firm's 20 contracts at the time they are purchased.

In buying currency futures, a firm does not actually make a payment equal to the value or price of the contract. Instead, the firm pays its broker a hedge or trade account margin, which is refunded to the firm at the time its contract is sold. This margin varies from broker to broker and also over time for a given broker, based on the volatility of the exchange rate. The margin amount may be posted in cash, a bank letter of credit, or in short-term U.S. Treasury instruments. There is also a brokerage fee per contract which becomes due when the contract is sold. For our example we assume that the trade margin per contract for marks is $2000 and that the brokerage fee per contract is $100. Furthermore, we assume that the total margin outlay of $40,000 ($2,000 × 20 contracts) is posted using a letter of credit so there is no opportunity cost associated with the transaction.

To estimate the profit or loss associated with currency futures, it is necessary to estimate the sell price for a contract in six months. In doing this, we assume a perfect hedge. A perfect hedge means that the difference between the spot price (exchange rate expressed in U.S. dollars per deutsche mark) and the futures price at the time the contract is sold is identical to the difference between the two prices at the time the contract was bought. The spot price for marks is computed by taking the reciprocal of the exchange rate, expressed in deutsche marks per U.S. dollar. The difference between the spot price for marks and the buy price for the futures contract on January 3 is -0.0082 (i.e., [1/2.4580] – 0.4150). To estimate the sell price in six months, we simply add the absolute value of this difference (0.0082) to the spot price for marks (the reciprocal of the exchange rate quoted in deutsche marks per U.S. dollar). The estimated sell price in six months is used to estimate the value of the 20 contracts at that time (estimated sell price × 125,000 marks per contract × 20 contracts). The spot prices, estimated sell prices, and estimated total contract values in six months for each planning case are provided in Table 8–9.

The estimated profit (or loss) associated with the purchase of currency futures is determined by subtracting the estimated total value of the contracts (180 days later) from the original purchase price (i.e., $1,037,760). The brokerage fee of $2,000 ($100 per contract times 20 contracts) is subtracted from (or added to) the estimated profit (or loss) to obtain an adjusted profit (or loss) estimate.[10,11] The initial and adjusted profit (or loss) estimates are provided in Table 8–9 for each planning case. Lastly, the actual cost to the buying firm is estimated by subtracting (or adding) the adjusted profit (or loss) estimate from (or to) the cost of the marks in six months. The results for each planning case are presented in the final row of Table 8–9.

Table 8–10 compares the actual costs for the three cases with the costs for the baseline approach.

TABLE 8–9
Summary of Required Calculations for Each Planning Scenario

	Buying Power of the U.S. Dollar		
	Increases 10 Percent (2.2122 DM/$1)	Remains Unchanged (2.4580 DM/$1)	Decreases 10 Percent (2.7038 DM/$1)
Spot price	$0.4520	$0.4068	$0.3699
Estimated sell price	0.4602	0.4150	0.3781
Estimated total value	1,150,500.00	1,037,500.00	945,250.00
Estimated profit (loss)	113,000.00	0.00	(92,250.00)
Adjusted profit (loss)	111,000.00	(2,000.00)	(94,250.00)
Cost of marks	1,093,120. 00	983,808.00	894,371.00
Estimated actual cost	982,120.00	985,808.00	988,621.00

In comparison to the baseline approach, futures save the buyer money in the event the buying power of the U.S. dollar declines, and result in an opportunity loss for the buyer in the event that the buying power of the U.S. dollar rises. Depending upon changes in the difference between the spot and futures prices at the time the contracts are bought and when they are sold, the actual cost to the buyer can vary for each planning case. A firm that decides to hedge its risk may or may not come

TABLE 8–10
Cost in U.S. Dollars for Three Planning Scenarios

	Buying Power of the U.S. Dollar		
Strategy	Decreases 10 Percent (2.2122 DM/$1)	Remains Unchanged (2.4580 DM/$1)	Increases 10 Percent (2.7038 DM/$1)
Currency futures	$ 982,120	$ 985,808	$988,621
Baseline approach	1,093,120	983,808	894,371

out ahead if it uses currency futures instead of forward buying to hedge (see Table 8-11). While the forward market provides absolute price certainty to the buying firm, the futures market offers the option of selling the contracts early in the event the buying power of the U.S. dollar is rising steadily. Such an action would minimize the loss sustained and thus reduce the total cost to the buyer from what it would have been otherwise.

Contractual Agreements (risk sharing)

One method of minimizing the risk to both the buyer and the supplier of an adverse exchange-rate movement is to use a risk-sharing contract. One common type of risk-sharing contract (Option in Table 8-11) stipulates that exchange-rate losses (or gains) are to be equally shared by both parties. Thus if the buying power of the U.S. dollar decreases, the loss incurred by the buyer computed with respect to the original contract price (expressed in U.S. dollars) is split in half and subtracted from the current (higher) cost of the needed quantity of foreign currency. If the buying power of the U.S. dollar increases, the gain experienced by the buyer is divided in two and added to the current (lower) cost of the needed quantity of foreign currency.

Another type of risk-sharing contract employs an "exchange-rate window." The window is defined as plus or minus some percentage movement in the exchange rate. As long as the exchange rate varies within the window, no adjustments to price are made. If the exchange rate moves outside the window, the price is adjusted. The adjustment process might require renegotiation of the contract price or it might involve a risk-sharing formula (e.g., losses or gains outside of the window are equally shared by both parties).

The results of applying the two risk-sharing options to the sample problem are presented in Table 8-11 and compared with the baseline approach of paying in the supplier's currency with no protection. The exchange-rate window option (Option 2) uses a window of plus or minus 5 percent; losses or gains outside the window are equally shared.

If the buying power of the U.S. dollar decreases by 10 percent, the buyer realizes a savings of $54,656 using Option 1 and $28,766 using

TABLE 8–11
Risk Sharing: Cost in U.S. Dollars for Three Planning Cases

	Buying Power of the U.S. Dollar		
Strategy	Decreases 10 Percent (2.2122 DM/$1)	Remains Unchanged (2.4580 DM/$1)	Increases 10 Percent (2.7038 DM/$1)
Option 1:	$1,038,464	$983,808	$939,090
Option 2*:	1,064,354	983,808	915,666
Baseline:	1,093,120	983,808	894,371

* For a 5 percent decrease in buying power, the exchange rate is 2.3351 DM/$1; for a 5 percent increase, the exchange rate is 2.5809 DM/$1.

Option 2. If the buying power of the U.S. dollar increases by 10 percent, the buyer incurs a loss of $44,719 under Option 1 and $21,295 under Option 2.

A risk-sharing contract may be requested by the supplier when payment is in U.S. dollars. The adjusted future cost to the buyer in this situation is identical to the adjusted future cost when payment is in the supplier's currency and risk sharing is used. This encompasses risk sharing done in conjunction with payments in both the supplier's and the buyer's currencies.

DISCUSSION

Table 8–12 provides a summary of the costs associated with the micro-level exchange-rate management strategies for the three planning cases. An examination of these results suggests the potential benefits of incorporating exchange-rate forecast information into the decision process. For example, if a buyer has forecast information that strongly indicates the buying power of the U.S. dollar is going to decrease over the time horizon of interest, it would be advantageous to pay in U.S. dollars. However, if the supplier raises the U.S. dollar price to incorporate a substantial risk premium or insists in payment in his own currency, the

TABLE 8–12
Comparison of Approaches: Cost in U.S. Dollars for Three
Planning Cases

	Buying Power of the U.S. Dollar		
Strategy	Decreases 10 Percent (2.2122 DM/$1)	Remains Unchanged (2.4580 DM/$1)	Increases 10 Percent (2.7038 DM/$1)
Pay in U.S. dollars	$ 983,808	$ 983,808	$ 983,808
Buying forward	1,000,207	1,000,207	1,000,207
Futures	982,120	985,808	988,621
Risk sharing option 1	1,038,464	983,808	939,090
Risk sharing option 2	1,064,354	983,808	915,666
Baseline	1,093,120	983,808	894,371

buyer would be well advised to pay in the supplier's currency and hedge his risk using currency futures or by buying forward. On the other hand, if the buyer anticipates that the value of the U.S. dollar is likely to fluctuate minimally, he might want to use risk-sharing Option 1 and make payment in whatever currency (buyer's or supplier's) the supplier prefers.

What if an exchange-rate forecast strongly suggests that the buying power of the dollar is going to rise? In this situation, the buying firm might want to pay in the supplier's currency with either no protection or with the limited protection afforded by risk-sharing Option 2. In the event that the supplier insists on payment in U.S. dollars, the buyer should negotiate for the inclusion of a risk-sharing clause in the contract.

The actual exchange rate in existence six months from January 3, 1986 was 2.1765 DM/$1. Thus, the buying power of the U.S. dollar deteriorated 11.5 percent! If on January 3 the buyer had anticipated a decrease in the value of the U.S. dollar against the German mark and had successfully negotiated payment in U.S. dollars without an added-risk surcharge, the outcome would have been quite satisfactory. If the supplier had insisted on payment in German marks and if the buyer had hedged his risk using currency futures (estimated cost is $980,870 using the settlement price for July 2, 1986), he would have achieved an even lower effective cost. If the buyer had hedged by buying forward (estimated cost is $1,000,207), his effective cost would still have been lower than the costs associated with risk sharing or the baseline approach.

LESSONS FOR MANUFACTURING

The analyses present in this chapter were intended to familiarize materials managers with the mechanics of macro- and micro-level exchange rate management techniques and demonstrate the potentially significant benefits of using exchange-rate management strategies in international purchasing.

It is clear that exchange-rate considerations are an important facet of international purchasing. The astute purchasing manager can save his or her company a great deal of money and make that company's products more competitive in the global marketplace through the wise use of exchange rate information.

APPENDIX A

Ignoring transaction costs (e.g., broker's fees) and daily settlements to market, and assuming a fully hedged position (i.e., the needed quantity of foreign exchange is exactly equal to the standard contract size) the hedger's loss (or profit) on the actual purchase of foreign exchange is exactly compensated for by an equivalent profit (or loss) on his futures dealing only if the difference between the spot price (actual price for foreign exchange) and the futures price is identical at the opening date (i.e., the date at which the futures contract is purchased) and the closing date (i.e., the date at which the futures contract is sold and the foreign exchange is purchased) of the set of transactions. The difference between the spot price and the futures price is called the basis (i.e., basis = cash price − futures price). (Usually the price of a futures contract is higher than the cash price of the foreign exchange with the difference representing the cost of holding the foreign exchange to the futures' month.)

If the basis is the same at the opening and closing dates of the set of transactions, the hedge is perfect and the hedger neither loses nor gains on balance. However since the two prices (spot and futures) rarely move precisely in step with each other, the hedger is usually left with a balance of profit or loss. If the absolute value of the basis increases between the opening and closing dates of the set of transactions, a profit results; if it decreases, a loss results. Uncertainty regarding the movement of the two prices is called "basis risk." In reality, the benefit of futures hedging is that you avoid the major risk of fluctuations in the price of the foreign exchange for the much smaller risk involved in fluctuations between the spot price and futures price of the foreign exchange. This residual risk is only a fraction of the risk you face without a hedge.

ENDNOTES

1. All of the theories and practical applications presented in this chapter were developed jointly with Dr. Shawnee K. Vickery of Michigan State University over several years of research.

2. J. Carter, R. Monczka, R. Narasimhan, and S. Vickery "International Purchasing: Research Agenda for the 1980s and Beyond," *Proceedings of the Annual Meeting for the Decision Sciences Institute*, November 1986.

3. Joseph R. Carter, "International Countertrade," In *Guide to Purchasing*, volume IV. (Tempe, Ariz. National Association of Purchasing Management 1987), pp. 1–15.

4. Stephan H. Goodman, "A Review of Foreign Exchange Rate Forecasting Techniques: Implications for Business and Policy," *Journal of Finance* vol. 34, no. 2, Feb. 1979, pp. 415–27.

5. Donald W. Dobler, Lamar Lee, Jr., and David N. Burt, *Purchasing and Materials Management: Text and Cases*, 4th ed. (New York: McGraw-Hill, 1984), p. 285.

6. The inventory holding cost used was 25 percent of the average inventory value per year. This resulted in holding costs of $10,621.60 and $34,509.10 for the monthly purchase plan and the volume timing purchase plan, respectively.

7. Conoco later confirmed that during 1986 this Japanese industry group claimed that collectively they lost approximately $2 billion.

8. It should be noted that the contract price could be higher because the supplier might incorporate a risk premium into the purchase price.

9. Since the foreign exchange is needed for payment in July, the trade is accomplished using September futures. This avoids the problem of "rolling over" a nearby contract (i.e., a June contract), so that transaction costs are minimized. Refer to the futures prices in The Wall Street Journal, January 6, 1986.

10. The procedure for estimating the adjusted profit or loss associated with the purchase of futures contracts is simplified for the purposes of illustration.

11. The application of the estimation procedure to the sample problem yields the following actual cost estimates for each planning case:

	Increase *10 percent*	*Remain* *Unchanged*	*Decrease* *10 percent*
Estimated actual cost	$982,621	$987,308	$991,345

INDEX

OTHER TITLES IN THE BUSINESS ONE IRWIN/APICS LIBRARY OF INTEGRATED RESOURCE MANAGEMENT

INTEGRATED DISTRIBUTION MANAGEMENT
Competing on Customer Service, Time, and Cost
Christopher Gopal and Harold Cypress

Manufacturing professionals who strive to maximize the efficiency of their companies' distribution systems will instantly recognize the wealth of knowledge this guide provides. Gopal and Cypress direct you toward satisfying internal and external customer expectations and provide strategies for improving efficiency and service using the powerful Integrating Link.

ISBN: 1-55623-578-X

INTEGRATED PRODUCTION AND INVENTORY MANAGEMENT
Revitalizing the Manufacturing Enterprise
Thomas E. Vollmann, William L. Berry, and D. Clay Whybark

Discover how to slash production and distribution costs by effectively monitoring inventory. *Integrated Production and Inventory Management* explains the inventory control processes that optimize customer service and improve purchasing forecasts and production schedules.

ISBN: 1-55623-604-2

EFFECTIVE PRODUCT DESIGN AND DEVELOPMENT
How to Cut Lead Time and Increase Customer Satisfaction
Stephen R. Rosenthal

Effective Product Design and Development will help you steer clear of long development delays by pointing out ways to detect design flaws early, and by showing how to empower the entire work team to recognize time-absorbing mistakes. You will discover how to shorten the cycle of new product design and development and turn time into a strategic competitive advantage.

ISBN: 1-55623-603-4

About APICS

APICS, the educational society for resource management, offers the resources professionals need to succeed in the manufacturing community. With more than 35 years of experience, 70,000 members, and 260 local chapters, APICS is recognized worldwide for setting the standards for professional education. The society offers a full range of courses, conferences, educational programs, certification processes, and materials developed under the direction of industry experts.

APICS offers everything members need to enhance their careers and increase their professional value. Benefits include:

- Two internationally recognized educational certification processes—Certified in Production and Inventory Management (CPIM) and Certified in Integrated Resource Management (CIRM), which provide immediate recognition in the field and enhance members' work-related knowledge and skills. The CPIM process focuses on depth of knowledge in the core areas of production and inventory management, while the CIRM process supplies a breadth of knowledge in 13 functional areas of the business enterprise.
- The APICS Educational Materials Catalog—a handy collection of courses, proceedings, reprints, training materials, videos, software, and books written by industry experts…many of which are available to members at substantial discounts.
- *APICS The Performance Advantage*—a monthly magazine that focuses on improving competitiveness, quality, and productivity.
- Specific industry groups (SIGs)—suborganizations that develop educational programs, offer accompanying materials, and provide valuable networking opportunities.
- A multitude of educational workshops, employment referral, insurance, a retirement plan, and more.

To join APICS, or for complete information on the many benefits and services of APICS membership, **call 1-800-444-2742** or **703-237-8344**. Use extension 297.